SpringerBriefs in Applied Sciences and Technology

SpringerBriefs present concise summaries of cutting-edge research and practical applications across a wide spectrum of fields. Featuring compact volumes of 50 to 125 pages, the series covers a range of content from professional to academic.

Typical publications can be:

- A timely report of state-of-the art methods
- An introduction to or a manual for the application of mathematical or computer techniques
- A bridge between new research results, as published in journal articles
- A snapshot of a hot or emerging topic
- An in-depth case study
- A presentation of core concepts that students must understand in order to make independent contributions

SpringerBriefs are characterized by fast, global electronic dissemination, standard publishing contracts, standardized manuscript preparation and formatting guidelines, and expedited production schedules.

On the one hand, **SpringerBriefs in Applied Sciences and Technology** are devoted to the publication of fundamentals and applications within the different classical engineering disciplines as well as in interdisciplinary fields that recently emerged between these areas. On the other hand, as the boundary separating fundamental research and applied technology is more and more dissolving, this series is particularly open to trans-disciplinary topics between fundamental science and engineering.

Indexed by EI-Compendex, SCOPUS and Springerlink.

Sesil Koutra

Energy in the Community

A Bet to Win

Springer

Sesil Koutra
Faculty of Architecture and Urban Planning
University of Mons
Mons, Belgium

ISSN 2191-530X ISSN 2191-5318 (electronic)
SpringerBriefs in Applied Sciences and Technology
ISBN 978-3-032-27147-1 ISBN 978-3-032-27148-8 (eBook)
https://doi.org/10.1007/978-3-032-27148-8

This Springer imprint is published by the registered company Springer Nature Switzerland AG
The registered company address is: Gewerbestrasse 11, 6330 Cham, Switzerland

To Zoé and Salvatore….

Acknowledgments

Special thanks and acknowledgments to my Springer editors for the opportunity to publish this work and make our research visible to all!

I want to express my deepest gratitude to the coordinators and colleagues of the ILES (Integrated Local Energy Systems)[1] research project for their insights and thoughtful feedback. This book is motivated by addressing local challenges related to global concerns about climate change, the energy transition, and decarbonization, driven by social equity and justice. Special thanks to the UMONS team, and especially Prof. Marc Frère, and my beloved PhD student, Mr. Chadi Mahfoud, for their support and insightful inputs.

On a personal note, I am deeply grateful to my family, Zoé, Salvatore, and my parents for their unwavering encouragement throughout this journey. I am grateful to you; your love and support are invaluable and a pillar that makes the impossible possible!

Thank you all!Competing InterestsThe author has no conflicts of interest to declare that are relevant to the content of this book.

[1] For more information: https://www.europarl.europa.eu/factsheets/fr/sheet/214/fonds-pour-une-transition-juste, Accessed: 02/06/2025

Contents

About the Author

Sesil Koutra is an engineer of urban planning and design and research and a teaching associate at the Faculty of Architecture and Urban Planning of the University of Mons (Belgium). After her studies in Greece and a professional roadmap in urban studies, she came to Belgium, where she developed a research activity (RESIZED—Research Excellence on Solutions and Implementation of Zero Energy Districts, FP7 project, https://cordis.europa.eu/project/id/621408/reporting) on the energy transition at urban scales, pursued by postdoctoral research on positive energy patterns. Since 2023, her research interests have been focused on energy communities and the challenges in Belgian territories. She has been a visiting researcher at various research institutions, including TU Delft and the University of the Basque Country, where she has been involved in PhD co-supervision and in contributions to European and national projects.

Key Contributor

Chadi Mahfoud PhD Candidate, Faculty of Architecture and Urban Planning, University of Mons, 7000, Mons.

Abbreviations

CEC	Citizen Energy Community
CETWA	Communautés d'Énergie et Technologie WAllonnes
CHP	Combined heat and power
CSC	Collective self-consumption
CWaPE	Commission Wallonne Pour Energie
DR	Demand response
EC	Energy community
EMD	Electricity Market Design
EPBD	Energy Performance Building Directive
EU	European Union
EWH	Elektrizitätswerk Hindelang
GHG	Greenhouse gas emissions
IDETA	Intercommunale de Développement Économique des Arrondissements de Tournai, d'Ath et de Communes Avoisinantes
IEMD	Internal Electricity Market Directive
ILES	Integrated Local Energy Systems
LEC	Local energy communities
MS	Member States
NIMBY	Not in My Back Yard
P2P	Peer-to-peer (trading)
PPAs	Power Purchase Agreements
PV	Photovoltaics
R&D	Research and development
REC	Renewable Energy Community
RED	Renewable Energy Directive
RES	Renewable energy sources
RSC	Renewable self-consumers
SME	Small-medium enterprises
SPOC	Single point of contact
UK	United Kingdom

Chapter 1
Introduction

1.1 Contextualization of Research: Description of the Problem

With the onset of the Industrial Revolution, CO_2 levels triggered an unprecedented rise in temperatures, sparking debates over the energy transition, and the debates on decentralization of systems have risen (Alanne and Saari 2006; Berka and Dreyfus 2021; Mulugetta et al. 2010). Energy communities are expected to significantly contribute to the decarbonization target by 2050, supporting democratization and enabling citizens to actively participate in decision-making through social innovation action plans and learning processes, which are expected to transform the European energy landscape (Nikolic et al. 2025).

Collective benefits of equitable energy production and distribution found their roots in 1970s within the "soft energy paths" developments (Lovins 1979) supported by bottom-up governmental structures ((Strachan et al. 2015; Van Veelen and Van der Horst 2018)). Importance and recent focus are given to societal and environmental aspects of energy justice (Sovacool et al. 2017), and the combat against energy poverty as a stark and alarming reality, recently defined in the EU legislation and the Energy Performance of Buildings Directive (EPBD) recast (European Parliament 2024) as the "lack of accessibility to essential energy services for basic needs responding to living and health standards."

The European engagements comprise strategic pillars to achieve a clean transition, while preserving equity and accessibility for all Europeans, as confirmed by the legislative package "Clean Energy for all Europeans," adopted in 2019 (European Commission 2019), for a secure, sustainable, competitive, and primarily affordable energy supply and accessibility. Through this lens, the European Commission identified trust in integrated markets as a means to promote energy efficiency by initiating community-driven incentives, emphasizing the importance of self-consumption and renewable energy, and reinforcing the use of renewable energy to fight energy

S. Koutra, *Energy in the Community*, SpringerBriefs in Applied Sciences and Technology, https://doi.org/10.1007/978-3-032-27148-8_1

poverty (Lowitzsch et al. 2023). Keeping in mind the overarching objectives of Energy Transition Fund Mechanisms (European Union 2021), Energy Communities (ECs) strike a promising, inclusive, scalable, and replicable model characterized by innovative structures of participation and entail the EC implication into the modern markets, e.g., Lowitzsch et al. (2023).

Nevertheless, this transformation requires the phase-out of conventional business models and dominant technologically based approaches (Sovacool 2016). It creates equal opportunities for users (e.g., Hisschemöller and Sioziou 2013), stakeholders, and prosumers (e.g., Butenko 2016), aligned with collective organizational models (Ruotsalainen et al. 2017), tailored to specific conditions and trialed to deliver social acceptance of technological advancements.

Given the pressing need to accelerate decarbonization and green transformation of energy and societal ecosystems, policymakers put forward the energy communities as a main contributor to (European Commission 2022):

1. Increase capacity and enable the use of renewable energies in decentralized energy infrastructure and networks.
2. Attract private investments and capital to promote actionable research on clean energy systems and reduce the requirements for public aid, e.g. Lowitzsch (2020).
3. Empower citizens and activate participation by placing the prosumers at the core of decision-making processes towards co-creation and co-design governance structures.

In this manifold context, energy communities, structured within shared values and a bedrock ideocracy around justice and security, are becoming increasingly consequential, bolstered by compact regulations, e.g., Renewable Energy Directive (REDII), Directive on Common Rules for the Internal Electricity Market (EU 2019/944), etc. and seen as promising actors for transformative ecosystems in sustainable and socially equitable standards, e.g., Matschoss et al. (2022), described as "honest brokers" for their "no one left behind" capacity, e.g., Martiskainen et al. (2018).

Given the spatial, legislative, but mainly social complexities, ECs are confronted with reciprocal challenges of insufficient knowledge and best practices, limitations of funding schemes englobed by complex administrative processes, failures of legislative frameworks, restrictions on transparency, and obsolete business models preventing the replicability of their models, explicitly developed in Sect. 2.4, where the traditional governance and technical models reflect the need for strategic dialogues and efficient collaborations to better frame the shared values and benefits.

1.2 Research Design, Scope, Motivation, and Limitations

With the introduction of key concepts of sustainability, environmental and societal transition, decarbonization, and decentralization, energy systems have emerged as a means to further uptake democratic, community-driven models. Defined as an innovative way to organize collective actions based on open and democratic provisions, ECs are a solution to energy transition challenges and an important key to the reinforcement of renewable energy use [21]. Belgium's approach to ECs is unique due to its federal structure (Flanders, Wallonia, and the Brussels region) with divided responsibilities, varying levels of progress, and convergences in the distinct regulatory frameworks.

The study is limited by its reliance on gray and scientific literature, as well as secondary data from desk research and on-site experiments. The author acknowledges the rapid evolution of the energy sector and the EC developments, emphasizing the need for further research and on-site experimentation. Future findings will incorporate insights from the ILES project (Sect. 1.2) and stakeholder experiences in Belgian cases and will explore additional perspectives to inform a more comprehensive and analytical design of ECs that will enable their scalability in Belgium and elsewhere.

1.2.1 ILES Research Project

This work is funded by the FTJ (Fonds de Transition Juste) project (CETWA)[1] coordinated by the University of Mons (UMONS). It is an initiative aimed at promoting sustainable and equitable energy transition in Wallonia and the decarbonization of the "Hainaut Heart." Focused on addressing the challenges associated with the shift to renewable energy, the project seeks to engage local actors in developing solutions that foster economic growth while minimizing environmental impacts. Through collaborative efforts, the project supports the implementation of green technologies, promotes energy efficiency, and creates new opportunities for *energy communities setup* and focused on the industrial sector, which remains the most polluting in Wallonia, and with particular emphasis on the districts of Tournai, Mons, and Charleroi.

This research is conducted within the framework of a €14 million portfolio led by UMONS, encompassing heating, cooling, and electricity and is part of Priority 6 of the ERDF-FTJ program "A Wallonia oriented toward a just transition, and more specifically, Measure 21 (Support for R&I actions—Development of research projects)"[2] in the field of technological solutions of energy communities and demonstrations in simulated and real environments. Within this broader initiative, *the work specifically investigates the drivers, barriers, and regulatory frameworks of energy communities in the Belgian context*, along with the development of digital tools and other relevant methodologies to enable the integration of decentralized

[1] For more information: https://www.europarl.europa.eu/factsheets/fr/sheet/214/fonds-pour-une-transition-juste, Accessed: 02/06/2025

[2] For more information: https://commission.europa.eu/funding-tenders/find-funding/eu-funding-programmes/european-regional-development-fund-erdf_en, Accessed: 02/06/2025

energy systems into energy sharing, coupling architectures tested under real-life conditions, and simulations that test user behavior.

Across all technologies, the R&D activities cover a TRL range from 3 to 7, including companies with diverse activities related to the development of energy technologies, dissemination processes, etc., aligned with the strategy of the IIS CETWA integrated. Along with a set of other initiatives that are active or under development, a key role in the energy transition and the socio-economic development in Wallonia is presented: a reflection on the use of sustainable and local energy resources, incorporating shared responsibility for energy production and consumption, and involving numerous stakeholders.

The project's overarching objectives align with the short-term, medium-term (2030), and long-term (2050) objectives, and with the creation of a favorable environment for energy initiatives by bringing together the public and private sectors. As a lever toward the clean and green energy transition, the concept of "Energy Communities" has been identified as a priority in the Wallonia region in terms of Research and Development, but also as a transition policy[3] composed of the following research areas:

1. Area 1: Collection, organization, and governance of energy data.
2. Area 2: Energy needs mapping and dynamic modeling. This axis, in which this work is developed, will enable the development of a methodological toolbox for the architectural and urban configurations of energy communities in Wallonia by conducting a spatial analysis of energy demand at different scales and a holistic diagnosis of energy flows aligned with typical prosumers' profiles. The technical and economic sizing of communities based on design tools will be adapted to urban, social, legislative, and other constraints, considering the trajectory of needs to develop a dynamic, forward-looking mapping study for needs assessment, tailored to urban, technological, political, economic, and societal contexts. The technologies to be developed, therefore, include geolocated, large-scale data, parametric, multi-criteria, dynamic methods, and statistical analyses. The developments will be based on the complete building stock of the Walloon Region, with a particular focus on Hainaut and on one or more case studies.
3. Area 3: Energy storage technologies (heat, electricity, and hydrogen).
4. Area 4: Use of low-temperature geothermal energy in energy communities.
5. Area 5: Combining energy technologies and testing on dynamic test benches.
6. Area 6: Tools for sizing and managing CERs deployed on multi-energy networks (tools and demonstrations in real-life ecosystems).

[3] For example, Conseil d'Etat du 28 mars 2025 a annulé l'article 4 de l'arrêté du Gouvernement du 17 mars 2023 relatif aux communautés d'énergie et au partage d'énergie, https://www.cwape.be/node/158, Accessed: 02/06/2025

1.2.2 Research Motivation and Limitations

The limits of centralized, fossil-fuel-based energy systems are widely acknowledged, making transformative design more necessary than ever. Aside from measures to invest in energy improvements, fossil fuel displacement has posed diverse challenges, including the need to support collective action at the local level and human-based solutions to facilitate the integration of sustainable energy technologies and practices. The concept of "energy community" has been described as a dynamic initiative in which people are at the core of decision-making processes to support effective policies and designs with significant social impacts, such as addressing energy poverty.

Community-based initiatives are important actors in the green transition and decarbonization, as well as in activating human-oriented projects. Recent research has investigated the factors influencing the stimulation of energy communities, but without systematic, homogeneous motivations and with significant ignorance of stakeholder synergies or innovation diffusion.

Initiated by the ILES research project, this study explores the drivers and gaps of energy communities projects in Belgian territories. Analytical and descriptive information on ongoing projects in the three federations is presented, highlighting the level of Belgian stakeholders' involvement and engagement, shaped by institutional, cultural, and political heterogeneity, spatial patterns, attitudes, and legal statuses. The activation of collective incentives and relevant mechanisms to promote and replicate energy communities in Belgium has shown promising results, although contributions remain fragmented and still underdeveloped, with numerous restrictions linked to conventional business models; the interplay with the social innovation scope should be further explored.

References

K. Alanne, A. Saari, Distributed energy generation and sustainable development. Renew. Sust. Energ. Rev. **10**, 539–558 (2006)

A. Berka, M. Dreyfus, Decentralisation and inclusivity in the energy sector: Preconditions, impacts and avenues for further research. Renew. Sust. Energ. Rev. **138**, 110663 (2021)

A. Butenko (2016). *User-centered innovation and regulation framework: Energy prosumers' market access in EU regulation* [Discussion paper]. https://papers.ssrn.com/sol3/papers.cfm?abstract_id=2797545

European Commission. *Clean energy for all Europeans package*. (2019). https://wayback.archive-it.org/12090/20241209144917/https://energy.ec.europa.eu/topics/energy-strategy/clean-energy-all-europeans-package_en

European Commission. Energy communities. Citizen-driven energy actions that contribute to advancing energy efficiency within communities. (2022). https://energy.ec.europa.eu/topics/markets-and-consumers/energy-consumers-and-prosumers/energy-communities_en

European Parliament. *Directive (EU) 2024/1275 of the European Parliament and of the Council of 24 April 2024 on the energy performance of buildings (recast)*. (2024). https://eur-lex.europa.eu/eli/dir/2024/1275/oj/eng

European Union. *Regulation (EU) 2021/1056 of the European Parliament and of the Council of 24 June 2021 establishing the Just Transition Fund.* (2021). https://eur-lex.europa.eu/legal-content/EN/TXT/?uri=CELEX%3A32021R1056&qid=1694610663421

M. Hisschemöller, I. Sioziou, Boundary organisations for resource mobilisation: enhancing citizens' involvement in the Dutch energy transition. Environ. Polit. **22**, 792 (2013)

A.B. Lovins, *Soft Energy Paths: Towards a Durable Peace* (HarperCollins, 1979)

J. Lowitzsch, Renewable energy communities under the 2019 European Clean Energy Package-governance for the energy clusters future? Renew. Sust. Energ. Rev. **122** (2020)

J. Lowitzsch, K. Kreutzer, J. George, C. Croonenbroeck, B. Breitschopf, Development prospects for energy communities in the EU identifying best practice and future opportunities using a morphological approach. Energy Policy **174**, 113414 (2023)

M. Martiskainen, E. Heiskanen, G. Speciale, Community energy initiatives to alleviate fuel poverty: the material politics of Energy Cafés. Local Environ. **23**, 20 (2018)

K. Matschoss, I. Mikkonen, L. Gynther, G. Koukoufikis, A. Uihlein, I. Murauskaite-Bull, Drawing policy insights from social innovation cases in the energy field. Energy Policy **161**, 112728 (2022)

Y. Mulugetta, T. Jackson, D. van der Horst, Carbon reduction at community scale. Energy Policy **38**, 7541–7545 (2010)

J. Nikolic, O. Clemmensen, M. Gerritsen, K. Sperling, H. J. Kooij. *COPPER project: energy communities: Inventory of barriers and opportunities and best practice*. (2025). https://dutpartnership.eu/sites/default/files/2025-11/COPPER%20ENERGY%20COMMUNITIES%20INVENTORY%20OF%20BARRIERS%20OPPORTUNITIES%20AND%20BEST%20PRACTICE.pdf

J. Ruotsalainen, J. Karjalainen, M. Child, S. Heinonen, Culture, values, lifestyles, and power in energy futures: a critical peer-to-peer vision for renewable energy. Energy Res. Soc. Sci. **34**, 231–239 (2017)

B.K. Sovacool, How long will it take? Conceptualizing the temporal dynamics of energy transition. Energy Res. Soc. Sci. **13**, 202–215 (2016)

B.K. Sovacool, M. Burke, L. Baker, C.K. Kotikalapudi, H. Wlokas, New frontiers and conceptual frameworks for energy justice. Energy Policy **105**, 677–691 (2017)

P.A. Strachan, R. Cowell, G. Ellis, F. Sherry-Brennan, D. Toke, Promoting community renewable energy in a corporate energy world. Sustain. Dev **23**, 96 (2015)

B. Van Veelen, D. Van der Horst, What is energy democracy? Connecting social science and political theory. Energy Res. Soc. Sci., 19–28 (2018)

Chapter 2
State-of-the-Art on Knowledge in Energy Communities

Using a systematic descriptive review, the origins of the "Energy Community" are unpacked, and the driving forces behind its conceptualization are outlined. Previous literature has covered and interconnected parameters on EC operation and relevant topics of renewable energy, sharing, and collectivity, such as self-consumption, sharing, flexibility, justice, cooperatives, and renewables (Fig. 2.1; Barabino et al. 2023; Bashi et al. 2023).

In a global panorama, prior research highlights the role of buildings and their loads in the mapping of EC understanding, basically focuses on scientific mathematical visualizations of the EC components, such as flexibility, e.g., Ponnaganti et al. (2023), or optimization viewpoints on energy demand response (DR) strategies to mitigate costs and efficiency, e.g., Belloni et al. (2025), or forecasting techniques for energy supplies and demand. Consequently, a spectrum of concepts in the scientific review (e.g., "community energy," "community solar," "community wind," etc.) attempt to analyze the specificities of the context, which remains inconsistent. Given the dual forces of reducing fossil fuel dependency and the shift to green and clean energy, the processes are usually critical, causing dilemmas that threaten justice and perpetuate insecurity, by changing distribution plans and affordability, leading to new kinds of vulnerabilities (Sovacool et al. 2019). In response, "Energy Community" aims to ensure equity in transitions and to broadly promote the decentralization of energy systems through collective, human-centric decision-making processes (Jenkins et al. 2021). Overall, various meanings of the community have emerged since its inception, emphasizing the importance of local actionable insights and distinguishing it from mega energy projects to promote sustainable development in broad terms (Taiwo and Tozer 2025).

Shaped by the motivations and driving forces behind tackling energy poverty, energy communities established connections between economic and environmental observations and represent a clean energy pathway toward an inclusive transition of collective actions initiated by individual motivations, e.g., environmental consciousness, level of education, awareness, willingness, income, shared values, and

S. Koutra, *Energy in the Community*, SpringerBriefs in Applied Sciences and Technology, https://doi.org/10.1007/978-3-032-27148-8_2

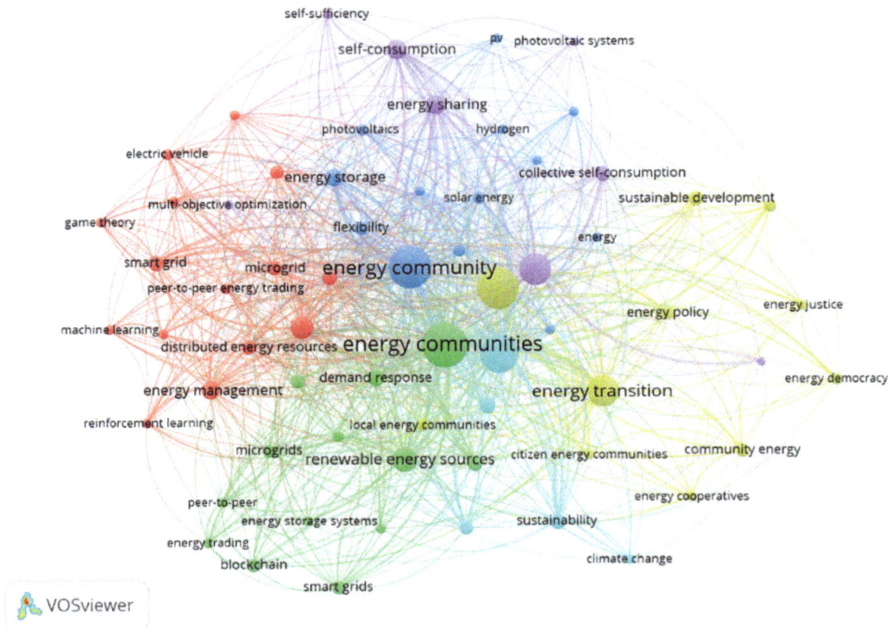

Fig. 2.1 Bibliometrics on energy communities. Keyword analysis. (©Scopus dataset, 2020–2026)

governance (Goedkoop et al. 2025). In collective terms, few studies explore citizen engagement in projects of renewable energy orientation, e.g., Bomberg and McEwen (2012) and Rogers et al. (2008). Scholars, such as Lindenberg and Steg (2007), have coupled motivations for common interests in collective action and sharing costs, or related literature focused on "collective prosumerism," e.g., Capellán-Pérez et al. (2018b), place-based, e.g., Renewable Energy Community.

Other meanings are cited in the literature, including technological dimensions (e.g., supply-side and demand-side), environmental motivations driven by European Directives transposed into national policies, and even energy autonomy ambitions. Broadly, the term is flexible, encompassing a plethora of configurations, types of actors, technologies, and business models (Iazzolino et al. 2022).

2.1 "Community" and "Energy": Historical Context and First Signs of Energy Communities

The earliest mention of "community" goes back to the philosopher *Aristotle*, in his book *Politics* (Mesaroş 2014), where humans are described as "political animals" (ζῷον πολιτικόν), who possess a naturally communitarian inclination to live in settlements, divided into districts and neighborhoods, and organized into "communities." Being a "political animal," humankind organized its first communities on survival motivations (e.g., food, shelter, etc.). Since then, the organization of "communities" has become a physical continuity along with the organization of sociopolitical interactions, including the energy sector (Krause 2019).

A modern design of the "Community" was envisioned by Ebenezer Howard, a renowned British urban planner from the nineteenth century, who started the garden city movement and gave a more holistic definition of the "Community" in his book *To-morrow*, as a response to the poor living conditions for low-income communities in industrial cities, like London. The goal of this urban model was to promote social equity through cooperative land ownership and provide equal access to green space and employment within the city. It was a blend of town and country benefits ("Town + Country = Garden City") where central city was for public buildings and zoned areas were for housing, industry, and agriculture (Cureton and Dunn 2014). The synthetic diagrams of the theoretical archetype aimed to capture the complexity of the "integrated community," minimizing commuting and increasing the social interactions in an "optimum" equilibrium with the green spaces. In this model, which represents the garden city communities, the concepts of sustainability and low-carbon strategies have already been developed.

Later on, in the 1970s, the term tracked when the alternative technology advocated "soft energy paths" to endorse the use of renewables within participatory and democratized actions' in wind energy projects in the mid-1980s to 1990s (Peeters et al. 2025), as triggered by the oil crisis and with obviously different characteristics; these shocks forced the reevaluation of energy accessibility and the reduction of fossil dependency. This way was a pioneer in mobilizing renewable energy from an economic and environmental perspective, making clearer the consequences of fossil fuel dependence (Gorroño-Albizu et al. 2019).

During these years, the first adoption of pilot projects has succeeded in an attractive ecosystem with broad public acceptance and funding support. As one of the leading countries on the EC projects, Denmark is an important European energy hub and has a decentralized governance system with a liberalized electricity market. At the same angle, Germany established ambitious targets and adopted specific measures to promote the energy transition (in 2016, there were more than 1000 registered energy—mainly wind power-based—initiatives in Germany), thanks to a well-established, alternative, and autonomous energy movements (Gancheva et al. 2018). Rapidly, the need for electrification has been prioritized and these first signs of communities have been integrated into larger perimeters, e.g., Verbong and Geels (2007), concerning more actors and depending on national contexts for their daily operations, e.g., Vernay et al. (2023).

Quickly, by embracing the green path of collectivity to incorporate new technologies to meet large-scale energy requirements and establish grid connections, energy communities gained popularity, attracting investments across Europe (Capellan-Pérez et al. 2018). Since that moment, renewable energy has become a key research area, with the most important outcome being the Danish project to generate energy in schools in the Tvind area featuring the world's most well-known turbine (GNL Wind Service 2025). In this line, the EC development of the renewable island of Samso in 1997, including 21 turbines in a circular process that empowered citizen participation, demonstrated feasibility and reduced energy dependency in 2005, and the population increased to 4000. From now on, Denmark actively supported wind energy through local initiatives and cooperatives through specific funding schemes, by increasing their growing impact and leading to cooperatives' development (Bauwens et al. 2016).

In the years following 2010, community ownership is estimated to hold more than 20% of wind power with commercialization scope (Menin 2021) and to reach more than 50% of PV production managed by communities in 2014. The initiative of the first ECs in Germany (1991) came after the Chernobyl nuclear accident, to liberalize the energy market, led by citizens, being recognized as ecological principles to *transform the grid and production, and install renewables* (RESCOOP.EU 2015). In 2009, this incentive became a cooperative to foster electricity sharing after the First World War, encouraging similar initiatives in rural areas, reaching 6000 in 1932, but without persistence due to the financial and bureaucratic barriers to distribution (Menin 2021).

From the 1990s onward, Great Britain and the Netherlands promoted renewable energy generation in urban areas, an essential milestone toward green collectivity and sharing (Mutani et al. 2019), with the first supporting mechanisms appearing in 1974 through the Renewable Energy Support Program for renewable energy development as an informative means toward community development and the PV inclusion in building scale. This initiative stemmed from the need for social inclusion, the problem of energy poverty, and independence from fossil fuels. Taking advantage of this, one of the first territories in England to experience ECs' bloom was Gorran Highlanes (St. Gorran, Cornwall), which promoted wind turbine facilities in a cooperative structure and for non-profit purposes (Moroni et al. 2019).

Since then, different "waves" of energy communities have appeared in the literature, initiated during the oil crisis, driven by incentives for self-sufficiency and efficiency in Denmark, and inspired by the UK's first wind farm. The second wave, later in the 1980s and 1990s, supported renewable energy use through relevant private investments (Padovan et al. 2019). Energy communities continue to develop due to a multitude of global crises, including pandemics, financial crises, natural disasters, and wars, as well as official policy trends and national regulations, namely the "Citizen" and "Renewable" concepts, which have gained significant popularity in several scientific documents and conceptual works.

France followed the marathon of ECs' development with the liberalization of the energy market in 2007. The concretization of its initiatives led by the national federation of Community Renewable Energy Projects, which enabled participatory investments and decision-making access, promoting several projects of energy decentralization in democratic patterns (Sebi and Vernay 2020), firstly introduced through the Energy Transition Law for Green Growth (2015) (Journal officiel de la République Française 2015) for installation, and operation of community projects and support their regulatory studies, leading to quadruple their number between 2014 and 2019 (Menin 2021). Similar initiatives have been experienced in Italy, especially from 2010 onward, due to the rapid increase in the installation of renewable plants and the bloom of cooperatives, especially in the northern part of the country, such as the pioneering Ewerk Prad to develop production in the Alto Adgani region of Italy and organize the demand response (Menin 2021). Since then, in the second half of the 2000s, rising cooperative experiences have emerged to provide locally sustainable community prototypes.

Another important country with valuable best practices and lessons learned on ECs' development is Spain, which is lagging behind other European countries due to operational barriers and social resistance. As explained by Capellán-Pérez et al. (Capellán-Pérez et al., 2018b), the absence of cultural context related to the renewable use, the intensive fossil use as the "as usual" business model, and the limited participatory empowerment are the main reasons for the slow shift toward clean transition. Going beyond this analysis, despite the liberalization of the electricity market in Spain in 1997 and the transposition of Directive 96/92/EC (European Parliament 1996) into Spanish regulations, the relevant market is still dominated by the private sector with profit-making purposes. To this, the consequences of the economic crisis (2008) led to dramatic reductions in electricity demand, combined with political pressures related to the RES exploitation (Belda and Pellicer 2016), a phenomenon that was reversed after the end of 2014 with the stimulation of synergies and active actor participation (Capellán-Pérez et al., 2018b), confirmed by the "Som Energia" cooperative supporting green initiatives in Seville and Almeria to cover more than 2500 households' electricity needs (Facchini 2021).

Other collective initiatives in several forms, schemes, and campaigns (Bomberg & McEwen, 2012) mobilize the term of energy community (or relevant) across Europe, with diverse meanings; for instance, in the Netherlands, the term is associated with social hubs (Dóci et al. 2015) with a strong commitment to energy efficiency and decarbonization strategies, while in the UK, important convergences have emerged in the cultures of communities between England (cooperative) and Scotland (community trust model) (Simcock et al. 2016).

Reporting more than 8000 projects, ECs' growth does not seem homogeneous across Europe, with more cases appearing in Germany and the Netherlands, followed by Austria, Switzerland, Denmark, France, Great Britain, Poland, Spain, and Sweden (Padovan et al. 2019); hence, the number of ECs in Europe remains approximate, e.g., Bauwens (2017). Caramizaru and Uihlein (2020) presented 24 cases from nine (9) European countries, considering broader targets, boundaries, and scales, ranging from large cooperatives to small citizen-organized energy cooperatives, going beyond technical dimensions to unpack social vulnerabilities and promote common benefits. The next wave has emerged since the global crisis period and the Fukushima nuclear disaster (2011), with rising interest in user-centric and energy-focused initiatives (De Moor 2013). Europe today has more than 9000 renewable energy cooperatives, with a significant presence across the continent and worldwide, including the United States, Australia, Canada, and New Zealand (Hoicka and MacArthur 2018).

Simcock et al. (2016), in their reports, introduced cases and lessons learned, acknowledged as leaders in the community-level energy sharing and management of international cases, e.g., Middelgrunden, as the largest community-led wind energy project, and Hvide Sande Wind Farm (Denmark), or the Bioenergy Village

Jühnde (Germany), Ecopower (Belgium), Brixton Energy (UK), or elsewhere in the world, e.g., Buan County (Korea). Looking across the distinct cases studied, specific structures reflect the institutional and political cultures and the shared responsibilities of local actors. Strong policies supporting the liberalization of energy markets, the decentralization of energy systems, and funding mechanisms were crucial drivers of ECs' development. Since their appearance, more than 1.250.000 citizens living in ECs or relevant structures are counted (Bierens and Skapoula 2021). In Europe, the earliest occurrence was in Italy in 1894, as a local cooperative in Chiavenna (Alpine), named "Società per l'Illuminazione Elettrica Cooperativa" developed around hydroelectric plants and a local distribution grid to provide electricity for the region of Valchiavenna (Cusa 2020).

Another remarkable project of early communities is the village of Jühnde in Lower-Saxony (Germany) as the first "bioenergy village," focusing on both heating and electricity production from renewables (biogas and biomass from agriculture and forestry (IEA 2022)), and based on self-sufficiency concepts and developed by the University of Göttingen in 2001 as an experiment for technological knowledge (Dóci and Vasileiadou 2015) and self-support design for 145 dwellings and 800 users, organized in a cooperative (IEA 2022). Later on, the study was framed by normative considerations for energy cost savings, aiming to provide 100% of electricity and at least 50% of heat needs, with interesting environmental engagement (Battisti 2020).

In 2019, the EU began the officialization of ECs' policy framework development for the operational promotion of the Energy Communities and estimates that by 2030, 17% of wind energy and 21% of solar energy of ECs will be human-centric oriented (Menin 2021).

These communities were essential for the deployment of the first supportive framework at the European and national levels. As part of Horizon 2020, Europe formalized the operational promotion of ECs, defining the legal differences between Renewable and Citizen structures (Menin 2021). Since 2010, the socio-political standards on community setup have supported the adoption of these citizen-led initiatives to promote the use of renewable energy and its activities (Roberts 2020).

2.2 Comprehensive Understanding of the Concepts: "Energy Community," "Community Energy," "Renewable," "Citizen," or Something Else?

Literature on collective prosumerism (Capellán-Pérez et al., 2018a; Walker and Devine-Wright 2008) has focused on related concepts, as the "community energy" (involvement or not renewables), e.g., Brummer (2018) and G. Walker and Devine-Wright (2008), "community solar," e.g., Hoffman and High-Pippert (2010), or "virtual power plants; peer-to-peer trading, and community-scale projects, e.g., Gui and MacGill (2018) with varied business models, governance arrangements, and technological achievements (Campos et al. 2020). Defined in EU Directives, ECs are legal entities enabling the generation, consumption, and collective energy management for shared benefits and values, empowering localized and decentralized systems by shifting from traditional business and "as usual" models to active prosumerism targeting both technological innovation and democratic structural governance, e.g., Köppl et al. (2022) and Neska and Kowlaska-Pyzalska (2022).

Collective initiatives in several forms, schemes, and campaigns mobilize the term "energy community" across Europe, with diverse meanings. In the energy context, the term "community" has often been conceived as a specific type of human relations characterized by participatory processes (Walker and Devine-Wright 2008), mapping dimensions of collaborative involvement (Bauwens et al. 2022). "Community" as a *process* refers to decision-making involving collaborative bodies from initial conception through common operations and benefits derived from collective activities, often characterized by social interactions and human-centric actions (Walker and Devine-Wright 2008). The term is usually presented as organizational or legal form (Rae and Bradley 2012), comprising shared values (Magnani and Osti 2016) to counter the polarization of climate pressures, often understood and characterized by horizontal governance and supporting systems mapping collective arrangements (Walker and Devine-Wright 2008) ranging from demand-supply to generation, distribution, and knowledge transfer, or even integration (Wouters et al. 2015), aggregation (Yazdani-Damavandi et al. 2017), and control (Mantovani et al. 2014) of electricity distribution.

Despite their polysemous and malleable terminology (e.g., Creamer et al. 2019), energy communities are conceptualized in diverse interests. More than 100 correlated terms appear, e.g., "community of renewable energies," "energy sustainable community," "energy cooperative," "integrated energy system" to define the phenomenon (Manso-Burgos et al. 2022), in technological viewpoints, e.g., Iazzolino et al. (2022). Across a spectrum of social arrangements and interactions, the term is overseen as a social innovation process, with active, empowered participation in decision-making and the distribution of benefits guided by the principle of energy justice. From this perspective, the term implies a high degree of voluntary involvement and transparency (Walker 2011). Energy communities are instrumentalized as human-centric activities that emphasize community participation (Braunholtz-Speight et al. 2021) and as a foothold in scientific research, while other definitions focus on peer-to-peer and market issues (Sousa et al. 2019).

Broadly speaking, energy communities are expressions of socio-technical phenomena and refer to energy-related initiatives led by a group of multi-motivated actors to promote joint investments in energy production and distribution for the common good and shared benefits and values. Generic notions of the specific use of infrastructure for collective energy management define the community in demand-side management and storage at local scales (Acosta et al. 2018; Parra et al. 2016); such a process highlights the importance of interactions in achieving low-carbon and energy transitions, shifting to decentralized local energy models (Brisbois 2020).

Alcayde et al. (2018) identified the most frequently used and acknowledged keywords in the field (see Table 2.1).

Table 2.1 Most commonly used keywords in ECs' definitions

System/method	Electric vehicles	Policy framework	Sustainability
Optimization	Smart grid	Energy policies	Social acceptance
Microgrid	Energy storage	Climate change	Community
Hybrid system	Optimization	Electricity	Energy transition
Hydrogen	Microgrids	Energy efficiency	User attitude
HOMER	District heating	Innovation	NIMBY
Energy storage	Demand response		Participation
Distributed generation	Energy efficiency		
Hybrid power system	Demand size management		
Smart grid	Vehicle-to-grid		
Energy management simulation	Electricity market		
Electrification	V2G		

Extensive work on the understanding of ECs' operational frame in the UK is proposed by Walker et al. (2010), driving the multiple benefits as developed in Middlmass and Parrish (2010) (environmental) or Joshi and Yenneti (2020) (social), despite the absence of a unified definition of the processes' dimensions to stimulate research, financial support, and participation empowerment (Rogers et al. 2008).

Due to this spectrum, the factors contributing to the community roadmap have been studied from different approaches (see Table 2.1), emphasizing the importance of collectivity and non-profit nature (e.g., Frieden et al. 2019) or unpacking the social (e.g., Wolsink 2012) or institutional (e.g., Wirth 2014) regards. No matter the definition's direction, the promotion of green rules in consumption and production is the core principle of ECs' design (Seyfang et al. 2013). Community has also been developed as a construct characteristic of spatial energy planning, embedded in localities and involving several interconnections of a technical nature, e.g., the sharing of resources by a group of people who together generate and store their electrical consumption (Schram et al. 2019).

Focusing on the settings of sustainable energy developments, scholars have ranged their interests in "community renewable energy" (e.g., Gordon Walker and Devine-Wright 2008), "energy communities" (e.g., Romero-Rubio and de Andrés Díaz 2015), "community solar energy" (e.g., Galindo Noguera et al. 2025), or "local cooperative energy systems" (e.g., Chilvers et al. 2021) expected to enact a key role in the transition toward clean strategies and systems' decentralization and actively engage citizens to the self-generation. A distinction in academic literature traces the convergences between "community energy" (extensively analyzed for over a decade) and "energy community," firstly introduced in the legislative package of 2016, including the dimension of "energy management" (Gui and MacGill 2018) related to the performance of diverse activities, as listed below:

1. *Generation/Production*: Collective production with possible surplus injected into the networks, e.g., Pons-Seres de Brauwer and Cohen (2020).
2. *Supplies*: Offer of aggregation services, or electricity production for commercial options, e.g., Reis et al. (2021).
3. *Distribution*: Core of the EC business model operation, the distribution of energy shared at local levels and in a specific perimeter, e.g., Caramizaru and Uihlein (2020).
4. *E-mobility*: Acting as a flexible, decentralized storage advantage to enable production and strengthen the grid stability and sufficiency.
5. Other activities developed in ECs levels include learning and training processes for the empowerment of active participation of prosumers, or educational campaigns for scaling up the best practices

Another literature stream, e.g., Bryant et al. (2018) and Hall and Roelich (2016), reviews the spectrum of ECs in line with their *financing mechanisms* for their activities and stakeholders, e.g., Brown et al. (2019), Mirzania et al. (2019), and Nolden et al. (2020), while others acknowledge the *social effects* of the EC structures on energy justice and democracy, e.g., Strachan et al. (2015) or the *environmental assets*, e.g., Berka and Creamer (2018).

2.3 Similarities and Convergencies on Citizen and Renewable Energy Community

A consensus is developed for energy self-consumption to bring out new terms in the literature (European Commission 2019a):

> [...] an open and voluntary participation is developed effectively controlled by the community members ('prosumers'), who are located in the proximity of renewable energy potential and are owners of the legal entity having as a principal goal to provide multiple benefits to its users.

The revised Internal Electricity Market Directive 2019/944 (IEMD) (European Commission 2019b) proposes the definition of *Citizen Energy* as a legal entity in which:

> [...] a voluntary, open, and effective participation of its members is generated, aiming to provide social and environmental rather than financial benefits and to engage them in the generation from RES and all the subsequent steps (distribution, consumption, storage) for the required energy services to the users.

According to this recast, CECs constitute a new type of legal entity due to the governance structure, with the specific purpose of providing services organized by citizens and relevant associations developed by them with energy-oriented activities and a strong emphasis on non-discriminatory access to electricity markets (Frieden et al. 2019)

.. whereas in the formal definition of Renewable Energy Directive (RED) is considered as (European Commission 2023b):

> [...] a legal entity based on open and voluntary participation, controlled by members or shareholders with a primary objective to provide environmental, economic, or social shared benefits, being engaged in providing distribution, supply, consumption, aggregation, storage, and other services **derived from renewables**.

As described in REDII, the REC should "be capable of remaining autonomous and located in the proximity of renewable energy potential; hence, proximity and autonomy remain interrelated for power generation and distribution." Defourny et al. (2020) primarily framed RECs as mutualistic organizations operating on the principle of reciprocity among diverse members, meaning that all participants retain control rights. The core mission for RECs remains the accessibility to clean energy plans derived from sustainable sources at competitive prices for all their members (Birchall and Simmons 2004). Table 2.2 provides a benchmark on the two terminologies based on European Directives' approaches.

From the above definitions, common features both take the form of legal entities based on open and voluntary participation with no profit purposes, but may differ in governance structures, and liabilities, e.g., Caramizaru and Uihlein (2020):

- *Geographical Limitation*: REC ties to the close proximity of renewable sources, aiming at the maximization of local hybrid production, whereas this is not specified in the case of CECs.
- *Energy Vector*: RECs focus exclusively on the renewable usage, while CECs may include different types, as natural or even fossil generators.
- *Activities*: CEC operates a priori for the electrical requirements of the community based on fossil sources. REC on the other side includes a rigid frame of renewable use and refers to a diverse range of activities in electricity and heating/cooling demand.
- *Participants*: No clear restrictions on participation are defined in CEC; engagements appear with large-scale energy actors with economic activities and

Table 2.2 Comparison of the "Renewable" and the "Citizen" Energy Community concepts according to EU Directives

Article 2(16) Recast "Renewable Energy Directive" on Renewable Energy Community (Directive 2023/2413) (European Commission 2023a)	Article 2(11) Recast Electricity Market Directive on "Citizen Energy Community" (Directive 2019/944) (European Commission 2019b)
A legal entity that	A legal entity that
Includes as shareholders, members, of which are natural persons, SMEs, or local authorities	Includes as shareholders, members, local authorities, or small enterprises
Has as primary scope the environmental, economic, and social common values and collective benefits for its members, where it operates, rather than financial profits	
Engage in generation, distribution, supply, storage, or other relevant activities *exclusively of renewable* energy carriers	Engage in generation, distribution, supply, storage, or other relevant activities of renewable *or fossil* energy carriers

Adapted from (Frieden et al. 2019)

decision-making power. REC involves a rigid participation limited to natural people, public authorities, and small/medium enterprises without profit, oriented toward social benefits (e.g., energy poverty).

- *Regulatory Treatment* (Vollmer and Fouquet 2021)*:* In terms of the regulatory framework, the key difference lies in the nature of the two directives, with the IEMD to recognize CECs formally as market sectors in the energy value chain, whereas RECs emerge from the promotion of renewable sources.

2.4 Who Is Involved in EC Projects? Mapping Stakeholders

Stakeholders are a key element of the EC setup and implementation, and their synergies are imperative to its successful operation. Stevens and Macharis (2009) define the stakeholders' ecosystem as a range of actors, who use the same systems or are influenced, directly or indirectly, by their use. Indeed, the stakeholder analysis is a matter of involved actors, and their nature, their role in the EC development, as well as the degree of their involvement in the process, with each situation being different, relevant to particularities and operational specifications, depending on the legal structure, the geographical and political context of their development, the scale of the project, the number of participants, the target audience, the goals, etc. (Heuninckx et al. 2022). Lode et al. (2019) identify potential EC stakeholders necessary for EC setup processes (Table 2.3).

In particular, contributors to EC development are involved in the whole value chain (Fig. 2.2). Key stakeholders are traditional planners (e.g., public authorities, landowners, planners, architects, and land managers), but also energy system providers, operators, and suppliers, or even facilitators, able to develop skills for balancing and aggregating local requirements e.g., Sareen et al. (2022), with complex synergies and governance structures across the diverse levels of implications socially oriented overarching measures during the conception and design. Citizens have a core role to be aligned with the technical requirements and adopt the proposed processes, which can increase the scalability and improvement of building stock in the studied cases.

Table 2.3 EC potential stakeholders

Stakeholder	Description
Initiator	Private person, private entity, or public authority (e.g., municipality)
EC participant	Member (private or public entity, consumer, or prosumer)
EC decision-making body	Private and/or entity responsible for the decision-making of EC deployment
Financing party	Public and/or private entity providing investment
Legal party	Public and/or private entity supporting EC with legal norms and relevant standards for operation and maintenance

Adapted from Lode et al. (2019)

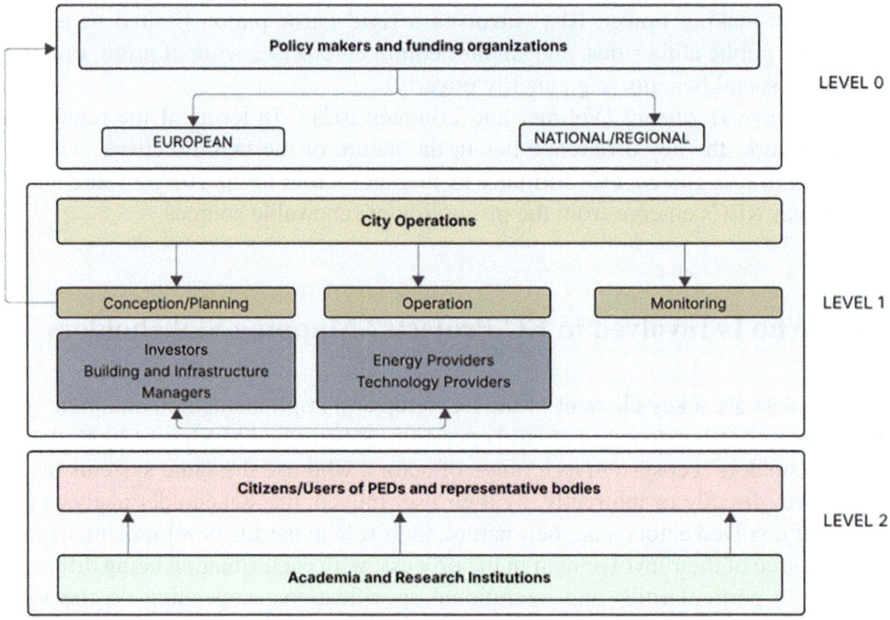

Fig. 2.2 Typical organization of stakeholders' levels in EC deployments. (©Koutra, 2026)

Level 0: Policymakers

Institutional policy and funding organizations get involved in regulatory and financial schemes for EC organization and promotion.

Level 1: Cities, Planning, Operation, and Monitoring Actors

- Cities play a leading role in the EC ecosystem and decision-making processes of holistic energy and urban planning. The city itself is the core operator in the decision-making of conception and implementation. The role of the cities' stakeholders is primordial to make the investments profitable, ensuring the financial schemes and the local operations for applicable design. Local governments turn attention to social innovations, attract private funding, and initiate new forms of collaboration and local governance to activate the local population.
- Energy, technology providers, and real estate developers dealing with energy management challenges.
- Investors for risk sharing, cooperation, and synergies, collective funding pipelines, and development of new and innovative business plans and models for management and secure the capacity of investments.
- Building and Infrastructure Managers: Their decision-making about building retrofitting and choices of energy technologies affect positive energy design.

Level 2: Citizens and Academia

- Citizens acting as prosumers.
- Academia, contributing to co-creation design with robust research enabling the EC deployment in long-term strategies and support in citizen engagement activi-

ties for the development of participation tools for social innovation actions seeking the alignment with European and national policies.

2.5 Addressing the Gaps on Energy Communities: Status Quo

Since their appearance, and explicit mention in European Directives, e.g., 2018/2001 (European Commission 2023b), 2019/944 (European Commission 2019b), etc., ECs are gaining attention in the growing body of literature, commonly acknowledged for their multiple positive outcomes, with a lack of evidence on the quantification of their social benefits and impacts, largely due to the heterogeneity of operational structures and the dynamic changes in energy markets (Blasch et al. 2021). Despite their pivotal role in the energy sector and multiple assets, ECs *have not yet reached their full potential due to multidimensional barriers*, e.g., Seyfang et al. (2013), including financial, legal, or societal issues, or lack of external supporting (e.g., networking, learning and awareness, funding, or technical knowledge), e.g., Kooij et al. (2018).

In practice, Member States have showed large spectrum in the effective implementation and performance, while in some of them the operations have not yet appeared due to differences in legal, organizational, or financial schemes, e.g., R. Hewitt et al. (2019), but also the plurality of actors and their different insights or conflicting interests, making the picture of the EC even more complex, e.g., Peeters et al. (2025).

Important barriers are also detected in the lack of *practical awareness and knowledge* among key stakeholders and policymakers involved, varying per country. Overall, in the list of barriers, scholars count basically the lack of clarified legal guidance with convergencies in respective countries and transposition in the national regulations, especially what constitutes the topics of energy sharing, as its practical framework is not clearly defined by the Clean Energy Package and hence is performed in numerous ways. In this lens, different efforts toward the clarification of this important topic for the ECs' operation topic are registered in the literature of current trends of EU Member States; for example, in the case of Austria, which has the most established and clarified governance structure allowing the energy sharing, or Denmark with the relevant regulations setup on electricity sharing within a single building (Nordic Energy Research 2023).

Overall, ECs are still a niche development with unexplored perspectives and great replication potential to bring out a clean transition, depending on technical skills and knowledge dissemination, e.g., Hufen and Koppenjan (2015), but with *significant gaps in non-technological entries* to enable processes of social innovation, learning, and value creation, which will allow decentralization and social inclusivity (Blasch et al. 2021). The way ECs are described and promoted usually reflects idealized views of justice and democracy, shifting toward decentralization, although fair and equitable practices do not meet energy poverty targets, which are rarely applied (Peeters et al. 2025).

The complexity of ECs replicability has raised much interest and various studies have appeared to investigate their aspects in a micro and macroenvironment, arguing the lacks on social issues, such as awareness and social acceptance and influence, justice, trust, or behavioral adaptation, or even managerial issues, as the varieties on business models and the investment capacity and attractiveness, but also environmental perspectives as mitigation of the carbon emissions, e.g., Neska and Kowlaska-Pyzalska (2022).

Due to the myriad interpretations, spectrums, contexts, configurations, and complexities in regulatory frameworks and business models, there are still gaps and ambiguities; an umbrella term to encompass the functionality, operation, governance, and technical coherence is still under exploration. In earlier research, institutional settings identify the technical support conditions and mechanisms for ECs' arrangements, often hindering social acceptance and behavioral change, thereby hindering scaling up in European territories. *What is primarily lacking is an understanding of the predominant key features, norms, and settings that emerge from differences in cultural, political, geographical, and societal contexts*, because of the past considerations that ECs vary in size, shape, boundaries, functionality, activities, and purposes; a typical case is an isolated island with no connection to the grid (Bauwens 2016).

Mapping the history of communities, their evolution, and patterns reveals the challenges posed by the lack of effective democratization and inclusiveness. Hanke et al. (2021) emphasize the male-dominated models, while Hannoset et al. (2019) point out the social pressure and resistance stemming from participatory processes, including possible bias, local interests, or ethics. Other authors, e.g., Renn et al. (2020), concluded that the bottom-up approaches should not be understood in their literal sense of justice, but as a pillar, where the different decisions are gravitated. Nonetheless, citizen incentives, energy-related, hinder *difficulties in reaching inclusivity* within actual business models, while scholars, e.g., Bielig et al. (2022), confirm the need for collaborative and multilevel actions. The need for extensive citizen engagement in more democratic energy landscapes claims the importance of bottom-up initiatives in models of collective self-consumption or peer-to-peer trading, going beyond previous technical contributions and optimization-based approaches developed for energy management and cost minimization, e.g., Chicco and Mancarella (2009).

Despite this, the mapping of the operational dimensions, the impetus of EC regulations, and an all-encompassing understanding of their functioning remain underdeveloped (Peeters et al. 2025). Generally speaking, and depending on the country, the main barriers to EC development are summarized as *administrative and legislative complexities*, difficulties in *financial schemes and trust*, *lack of incentives*, *transparency*, *and motivations* as missing pieces of this puzzle, where the communities emerge.

In order to scale the technical intervention, ECs should deal with exogenous and endogenous barriers of their operational systems, including misunderstandings and strong dependence of national policies, usually partially interpretated, e.g., Herbes et al. (2017) and Mirzania et al. (2019), or lack of technical designs related to installations and systems' failures, e.g., the grid (e.g., Bolinger et al. 2001), or even restricted trust and behavioral barriers or lack of knowledge and willingness for participation, e.g., Hoffman and High-Pippert (2010). So far, the national transpositions of EU face two-front challenges (analyzed in Chap. 3): (a) ensuring compliance with governance requirements of ECs and (b) practical complexities on energy sharing, e.g., Sokolowski (2020).

Another important gap to ensure the ECs optimal design from a technical point is the *common acceptance and technological adoption*, and the lack of clarified setting-up engagement and development of strong links from the EC project's early conception to enhance the knowledge dissemination, and build trust, fitted to local requirements, ensuring smooth planning processes and more successful and efficient implementation (Heuninckx et al. 2022). Prior research has already demonstrated that the early engagement of local stakeholders reflects lower costs, fewer delays, and less uncertainty, e.g., DTI (2003). To the author's best knowledge, tools for stakeholders' engagement in EC processes are limited, basically developed along with decision-making applications around policy planning, e.g., Randolph and Bauer (2005). Most research on stakeholders' motivations for EC projects, and influencing factors of behavioral intentions, personal, and social norms needs to be further developed, e.g., Huijts et al. (2012), or relevant attitudes on environmental concerns, renewable energy use, etc. (Bauwens 2019).

References

C. Acosta, M. Ortega, T. Bunsen, B. Koirala, A. Ghorbani, Facilitating energy transition through energy commons: An application of socio-ecological systems framework for integrated community energy systems. Sustainability **10** (2018)

A. Alcayde, F. Montoya, R. Banos, A. Perea-Moreno, F. Manzano-Agugliaro, Analysis of research topics and scientific collaborations in renewable energy using community detection. Sustainability (2018)

E. Barabino, D. Fioriti, E. Guerrazzi, I. Mariuzzo, D. Poli, M. Raugi, E. Razaei, E. Schito, D. Tholopulos, Energy communities: A review on trends, energy system modelling, business models, and optimisation objectives. Sustain. Energy Grids Netw. **36** (2023)

M.H. Bashi, L. De Tommasi, A. Le Cam, L. Sánchez Relaño, P. Lyons, J. Mundó, I. Pandelieva-Dimova, H. Schapp, K. Loth-Babut, C. Egger, M. Camps, B. Cassidy, G. Angelov, C.E. Stancioff, A review and mapping exercise of energy community regulatory challenges in European member states based on a survey of collective energy actors. Renew. Sust. Energ. Rev. (2023)

R. Battisti. *Autoconsumo collettivo e comunità energetiche: Sviluppi futuri e raccomandazioni.* (2020). https://www.pvp4grid.eu/wp-content/uploads/2020/04/D4.3-National-Policy-Paper-Italy.pdf

T. Bauwens, Explaining the diversity of motivations behind community renewable energy. Energy Policy **93**, 278–290 (2016)

T. Bauwens, Polycentric governance approaches for a low-carbon transition: the roles of community-based energy initiatives in enhancing the resilience of future energy systems, in *Complex Systems and Social Practices in Energy Transitions*, (2017)

T. Bauwens, Analyzing the determinants of the size of investments by community renewable energy members: Findings and policy implications from Flanders. Energy Policy **129**, 841–852 (2019)

T. Bauwens, B. Gotchev, L. Holstenkamp, What drives the development of community energy in Europe? The case of wind power cooperatives. Energy Res. Soc. Sci. **13**, 136–147 (2016)

T. Bauwens, D. Schraven, E. Drewing, J. Radtke, L. Holstenkamp, B. Gotchev, O. Yildiz, Conceptualizing community in energy systems: a systematic review of 183 definition. Renew. Sust. Energ. Rev. (2022)

M.S. Belda, S.V. Pellicer, *Rethinking Climate Change from Grassroots Social Innovation: Approaches* (International Congress of Development Studies, 2016)

E. Belloni, T. Ferrucci, D. Fioriti, A. Tumiati, D. Poli, Global diffusion and key features of energy communities with a main focus on building loads modelling and management: A review. Appl. Energy **399**, 126461 (2025)

A.L. Berka, E. Creamer, Taking stock of the local impacts of community owned renewable energy: A review and research agenda. Renew. Sust. Energ. Rev. **82**, 3400–3419 (2018)

M. Bielig, C. Kacperski, F. Kutzner, S. Klingert, Evidence behind the narrative: critically reviewing the social impact of energy communities in Europe. Energy Res. Soc. Sci. **94** (2022)

T. Bierens, A. Skapoula, Energy communities: the hidden gems of the EU energy transition. Energy Transition-The Global Energiewende. (2021) https://energytransition.org/2021/10/energy-communities-the-hidden-gems-of-the-eu-energy-transition/

J. Birchall, R. Simmons, What motivates members to participate in co-operative and mutual businesses? Annals Public Coopera. Econ. **75**, 465–495 (2004)

J. Blasch, N.M. van der Grijp, D. Petrovics, J. Palm, N. Bocken, S.J. Darby, J. Barnes, P. Hansen, T. Kamin, U. Golob, M. Andor, S. Sommer, A. Nicita, M. Musolino, M. Mlinarič, New clean energy communities in polycentric settings: four avenues for future research. Energy Res Social Sci **82** (2021)

M. Bolinger, R. Wiser, L. Milford, M. Stoddard, K. Porter, States emerge as clean energy investors: a review of state support for renewable energy. Electr. J. **14**, 82–95 (2001)

E. Bomberg, N. McEwen, Mobilizing community energy. Energy Policy, 434–444 (2012)

T. Braunholtz-Speight, C. McLachlan, S. Mander, M. Hannon, J. Hardy, I. Cairns, M. Sharmina, E. Manderson, The long-term future for community energy in Great Britain: a co-created vision of a thriving sector and steps towards realising it. Energy Res. Soc. Sci. **78** (2021)

M.C. Brisbois, Shifting political power in an era of electricity decentralization. Rescaling, reorganization and battles for influences. Environ. Innov. Soc. Trans. **36**, 49–69 (2020)

D. Brown, S. Hall, M. Davis, Prosumers in the post subsidy era: an exploration of new prosumer business models in the UK. Energy Policy **135** (2019)

V. Brummer, Community energy-benefits and barriers: a comparative literature review of community energy in the UK, Germany and the USA. The benefits it provides for society and the barriers it faces. Renew. Sust. Energ. Rev. **94**, 187–196 (2018)

S.T. Bryant, K. Straker, C. Wrigley, The typologies of power: energy utility business models in an increasingly renewable sector. J. Clean. Prod. **195**, 1032–1046 (2018)

C. Campos, P.L. Guilherme, M.-G. Esther, G. Swantje, H. Stephen, H. Holstenkamp, Regulatory challenges and opportunities for collective renewable energy prosumers in the EU. Energy Policy **138**. Scopus (2020). https://doi.org/10.1016/j.enpol.2019.111212

I. Capellán-Pérez, A. Campos-Celador, J. Terés-Zubiaga, Is community energy really non-existent in post-socialist Europe? Examining recent trends in 16 countries. Energy Policy, 215–229 (2018a)

I. Capellán-Pérez, A. Campos-Celador, J. Terés-Zubiaga, Renewable energy cooperatives as an instrument towards the energy transition in Spain. Energy Policy **123**, 215–229 (2018b)

A. Caramizaru, A. Uihlein, *Energy Communities: An Overview of Energy and Social Innovation* (JRC Science for Policy Report, European Commission, 2020) https://publications.jrc.ec.europa.eu/repository/handle/JRC119433

G. Chicco, P. Mancarella, Distributed multi-generation: a comprehensive view. Renew. Sust. Energ. Rev. **13**, 535–551 (2009)

J. Chilvers, R. Bellamy, H. Pallett, T. Hargreaves, A systemic approach to mapping participation with low-carbon energy transitions. Nat. Energy **6**, 250–259 (2021)

E. Creamer, G. Taylor Aiken, B. Van Veelen, G. Walker, P. Devine-Wright, Community renewable energy: What does it do? Walker and Devine-Wright (2008) ten years on. Energy Res. Social Sci. **57** (2019)

P. Cureton, N. Dunn, *Future of Cities: A Visual History of the Future* (Bloomsbury, 2014)

E. Cusa, Sustainable development, active citizenship and energy communities. Orizzonti del Diritto Commerciale. (2020) https://www.diseade.unimib.it/sites/st02/files/Cusa%20 pubblicazioni/41.2020.comunita.energetiche.Rivista.ODC_.pdf

T. De Moor, *Homo Cooperans* (Institutions for collective action and the compassionate society, Universiteit Utrecht, 2013) https://dspace.library.uu.nl/handle/1874/349371

J. Defourny, M. Nyssens, O. Brolis, Testing social enterprise models across the world: Evidence from the 'International Comparative Social Enterprise Models (ICSEM)' project. Nonprofit Volunt. Sect. Q. **50**, 420 (2020)

G. Dóci, E. Vasileiadou, 'Let's do it ourselves'. Individual motivations for investing in renewables at community level. Renew. Sust. Energ. Rev. **49**, 41–50 (2015)

G. Dóci, E. Vasileiadou, A.C. Petersen, Exploring the transition potential of renewable energy communities. Futures **66**, 85–95 (2015)

DTI, *Energy White Paper. Our energy future-creating a low carbon economy* (Department for Transport, Department for Environment, Food and Rural Affairs, 2003) https://assets.publishing.service.gov.uk/media/5a7c1f5940f0b645ba3c6d4f/5761.pdf

European Commission. *Clean energy for all Europeans package*. (2019a). https://wayback.archive-it.org/12090/20241209144917/https://energy.ec.europa.eu/topics/energy-strategy/clean-energy-all-europeans-package_en

European Commission. *Directive (EU) 2019/944 of the European Parliament and of the Council of 5 June 2019 on common rules for the internal market for electricity and amending Directive 2012/27/EU*. (2019b). https://eur-lex.europa.eu/eli/dir/2019/944/oj/eng

European Commission. *Directive (EU) 2023/2413 of the European Parliament and of the Council of 18 October 2023 amending Directive (EU) 2018/2001, Regulation (EU) 2018/1999 and Directive 98/70/EC as regards the promotion of energy from renewable sources, and repealing Council Directive (EU) 2015/652*. (2023a). https://eur-lex.europa.eu/legal-content/EN/TXT/?uri=CELEX%3A32023L2413&qid=1699364355105

European Commission. (2023b). Renewable energy directive (RED II) directive (EU) 2018/2001 (recast) on the promotion of the use of energy from renewable sources.. https://www.europex.org/eulegislation/renewable-energy-energy-directive/

European Parliament. *Directive 96/92/EC of the European Parliament and of the Council of 19 December 1996 concerning common rules for the internal market in electricity*. (1996). https://eur-lex.europa.eu/eli/dir/1996/92/oj/eng

M. Facchini, Da Magliano Alpi a Biccari, le storie delle comunità energetiche in Italia. Altreconomia. (2021) https://altreconomia.it/da-magliano-alpi-a-biccari-le-storie-delle-comunita-energetiche-in-italia/

D. Frieden, A. Tuerk, J. Roberts, S. D'Herbemont, A. Gubina. *Collective Self-Consumption and Energy Communities: Overview of Emerging Regulatory Approaches in Europe. Compile: Integrating Community Power in Energy Islands*. (2019). https://www.rescoop.eu/uploads/rescoop/downloads/COMPILE_Collective_self-consumption_EU.pdf

A.L. Galindo Noguera, L.S. Mendoza Castellanos, S. Ardila Cruz, G. Osorio-Gomez, Key aspects and challenges for successful energy communities: a comparative analysis between Latin America and developed countries. Renew. Sust. Energ. Rev. **216**, 115687 (2025)

M. Gancheva, S. O'Brien, N. Crook, C. Monteiro, *Models of Local Energy Ownership and the Role of Local Energy Communities in Energy Transition in Europe* (European Committee of the Regions, 2018)

GNL Wind Service. *Tvind wind turbine*. (2025). https://gnlservice.dk/en/tvind-wind-turbine/

F. Goedkoop, L. Jans, G. Perlaviciute, K.R.S. Hamann, Inclusive community energy initiatives? Understanding involvement of different socio-demographic groups. Energy Res. Social Sci. **125** (2025)

L. Gorroño-Albizu, K. Sperling, S. Djorup, The past, present and uncertain future of community energy in Denmark: critically reviewing and conceptualizing citizen ownership. Energy Res. Social Sci. (2019)

E.M. Gui, I. MacGill, Typology of future clean energy communities: an exploratory structure, opportunities and challenges. Energy Res. Social Sci. **35**, 94 (2018)

S. Hall, K. Roelich, Business model innovation in electricity supply markets: the role of complex value in the United Kingdom. Energy Policy **92**, 286–298 (2016)

F. Hanke, R. Guyet, M. Feenstra, Do renewable energy communities deliver energy justice? Exploring insights from 71 European cases. Energy Res. Social Sci. **80** (2021)

A. Hannoset, L. Peeters, A. Tuerk, *Energy Communities in the EU Task Force Energy Communities* (European Commission, 2019)

C. Herbes, V. Brummer, J. Rognli, S. Blazejewski, N. Gericke, Responding to policy change: new business models for renewable energy cooperatives-barriers perceived by cooperatives' members. Energy Policy **109**, 82–95 (2017)

S. Heuninckx, G.T. Boveldt, C. Macharis, T. Coosemans, Stakeholder objectives for joining an energy community: Flemish case studies. Energy Policy **162**. Scopus (2022). https://doi.org/10.1016/j.enpol.2022.112808

R. Hewitt, N. Bradley, A.B. Compagnucci, C. Barlagne, A. Ceglarz, R. Cremades, M. McKeen, I.M. Otto, B. Slee, Social innovation in Community Energy in Europe: a review of the evidence. Front. Energy Res. (2019)

S.M. Hoffman, A. High-Pippert, From private lives to collective action: recruitment and participation incentives for a community energy program. Energy Policy **38**, 7567–7574 (2010)

C.E. Hoicka, J.L. MacArthur, From tip to toes: mapping community energy models in Canada and New Zealand. Energy Policy **121**, 162–174 (2018)

J.A. Hufen, J.F. Koppenjan, Local renewable energy cooperatives: revolution in disguise? Energy Sustain. Soc. **5** (2015)

N.M.A. Huijts, E.J.E. Molin, L. Steg, Psychological factors influencing sustainable energy technology acceptance: a review-based comprehensive framework. Renew. Sust. Energ. Rev. **16**, 525–531 (2012)

G. Iazzolino, N. Sorrentino, D. Menniti, A. Pinnarelli, L. De Carolis, L. Mendicino, Energy communities and key features emerged from business models review. Energy Policy (2022)

IEA. *Biogas in the society. Information from IEA Bioenergy. Task 37. Energy from biogas and landfill gas. The first bioenergy village in Juhnde/Germany. Energy self sufficiency with biogas*. (2022). https://task37.ieabioenergy.com/wp-content/uploads/sites/32/2022/03/biogas_village.pdf

E. Jenkins, B.K. Sovacool, N. Mouter, N. Hacking, M.K. Burns, D. McCauley, The methodologies, geographies and technologies of energy justice: a systematic and comprehensive review. Environ. Res. Lett. **16**, 043009 (2021)

G. Joshi, K. Yenneti, Community solar energy initiatives in India: a pathway for addressing energy poverty and sustainability? Energy Build. **210** (2020)

Journal officiel de la République Française. *Loi no 2015-992 du 17 aout 2015 relative à la transition énergétique pour la croissance verte*. (2015). https://cdn.climatepolicyradar.org/navigator/FRA/2015/law-no-2015-992-on-energy-transition-for-green-growth-energy-transition-law_ce0a35ebcc760af8c2fcaae5e7796360.pdf

H.J. Kooij, M. Oteman, S. Veenman, K. Sperling, D. Magnusson, J. Palm, F. Hvelplund, Between grassroots and treetops: Community power and institutional dependence in the renewable energy sector in Denmark, Sweden and The Netherlands. Energy Res. Soc. Sci. **37**, 52–64 (2018)

S. Köppl, E. Springmann, V. Regener, A. Weigand, *Enabling Business Models and Grid Stability: Case Studies from Germany. Energy Communities* (Academic Press, 2022), pp. 229–243

P. Krause. *Aristotle's "Political Animal". Discourses on Minerva* (2019, June 28). https://minervawisdom.com/2019/06/28/aristotles-political-animal/

S. Lindenberg, L. Steg, Normative, gain and hedonic goal frames guiding environmental behavior. J. Soc. Issues **63**, 117 (2007)

L. M. Lode, S. Heuninckx, C. A. Koronen. *ROLECS deliverable 4.3-stakeholder mapping and scenario building* [Deliverable, Project ROLECS] (2019)

N. Magnani, G. Osti, Does civil society matter? Challenges and strategies of grassroots initiatives in Italy's energy transition. Energy Res. Social Sci. **13**, 148 (2016)

A. Manso-Burgos, D. Ribo-Pérez, T. Gomez-Navarro, M. Alcazar-Ortega, Local energy communities modelling and optimization considering storage, demand configuration and sharing strategies: a case-study in Valencia (Spain). Energy Rep., 10395–10408 (2022)

G. Mantovani, G.T. Costanzo, M. Marinelli, L. Ferrarini, Experimental validation of energy resources integration in microgrids via distributed predictive control. **29**, 1018–1025 (2014)

S. Menin. *Regulating energy communities in Europe: From the origins to the EU Directives* [PhD Thesis, Universita'Degli Studi di Paova. Dipartimento di Scienze Economiche ed Aziendali 'M.Fanno']. (2021). https://thesis.unipd.it/retrieve/8568ddb5-e1ea-495b-9d62-823c72fe08ad/Menin_Simone.pdf

C. Mesaroş, Aristotle and animal mind. Procedia. Soc. Behav. Sci. **163**, 185–192 (2014). https://doi.org/10.1016/j.sbspro.2014.12.305

L. Middlmass, B.D. Parrish, Building capacity for low-carbon communities: The role of grassroots initiatives. Energy Policy, 7559–7566 (2010)

P. Mirzania, A. Ford, D. Andrews, G. Ofori, G. Maidment, The impact of policy changes: The opportunities of community renewable energy projects in the UK and the barriers they face. Energy Policy **129**, 1282–1296 (2019)

S. Moroni, V. Alberti, V. Antoniucci, A. Bisello, Energy communities in the transition to a low-carbon future: a taxonomical approach and some policy dilemmas. J. Environ. Manag. **236**, 45 (2019)

G. Mutani, V. Todeschi, S. Santantonio, M. Bazzino, Energy communities: A new energy policy at territorial scale, in *Conference: Energy for Sustainability International Conference 2019*, (Sustainable Future, Designing a, 2019)

E. Neska, A. Kowlaska-Pyzalska, Conceptual design of energy market topologies for communities and their practical applications in EU: a comparison of three case studies. Renew. Sust. Energ. Rev. (2022)

C. Nolden, J. Barnes, J. Nicholls, Community energy business model evolution: A review of solar photovoltaic developments in England. Renew. Sust. Energ. Rev. **122**, 109722 (2020)

Nordic Energy Research. (2023). Energy communities. https://pub.norden.org/nordicenergyresearch2023-03/netherlands.html

Padovan, D., Arrobio, O., Sciullo, A., Gilcrease, W., Gregg, J. S., Henfrey, T., Wierling, A., Schwanitz, V. J., Labanca, N., Dunlop, T., Alvarez Polo, L., & Candelise, C. (2019). *Collective Action Initiatives. Some Theoretical Perspectives and a Working Definition*

D. Parra, S.A. Norman, G.S. Walker, M. Gillott, Optimum community energy storage system for demand load shifting. Appl. Energy **174**, 130–143 (2016)

L. Peeters, L. Fernandez Lopez, C. Trompoukis, Addressing the Gaps in Understanding and Assessing Energy Communities. Energy Res. Social Sci. **127** (2025)

P. Ponnaganti, R. Sinha, R. Pillai, B. Bak-Jensen, Flexibility provisions through local energy communities: a review. *Next* Energy **1** (2023)

C. Pons-Seres de Brauwer, J.J. Cohen, Analysing the potential of citizen-financed community renewable energy to drive Europe's low-carbon energy transition. Renew. Sust. Energ. Rev. **133**, 110300 (2020)

C. Rae, F. Bradley, Energy autonomy in sustainable communities-a review of key issues. Renew. Sust. Energ. Rev. **16**, 6497–6506 (2012)

J. Randolph, M. Bauer, Improving environmental decision-making through collaborative methods. Rev. Policy Res., 168–191 (2005)

I. Reis, I. Gonçalves, M. Lopes, C. Henggeler Antunes, Business models for energy communities: a review of key issues and trends. Renew. Sust. Energ. Rev. (2021)

O. Renn, F. Ulmer, A. Deckert, *The Role of Public Participation in Energy Transitions* (Academic Press, 2020)
RESCOOP.EU. *The REScoop model.* (2015). https://www.rescoop.eu/the-rescoop-model
J. Roberts, Power to the people? Implications of the Clean Energy Package for the role of community ownership in Europe's energy transition. Rev. Eur. Comp. Int. Environ. Law **29**, 232–244 (2020)
J.C. Rogers, E.A. Simmons, I. Convery, A. Weatherall, Public perceptions of opportunities for community-based renewable energy projects. Energy Policy **36**, 4217–4226 (2008)
C. Romero-Rubio, J.R. de Andrés Díaz, Sustainable energy communities: A study contrasting Spain and Germany. Energy Policy **85**, 397–409 (2015)
S. Sareen, V. Albert-Seifried, L. Aelenei, F. Reda, G. Etminan, M.B. Andreucci, M. Kuzmic, N. Maas, O. Seco, P. Civiero, S. Gohari, M. Hukkalainen, H.M. Neumann, Ten questions concerning positive energy districts. Build. Environ **216**, 109017 (2022)
W. Schram, A. Louwen, I. Lampropoulos, W. Van Sark, Comparison of the greenhouse gas emission reduction potential of energy communities. Energies **12** (2019)
C. Sebi, A. Vernay, Community renewable energy in France: The state of development and the way forward. Energy Policy **147** (2020)
G. Seyfang, J.J. Park, A. Smith, A thousand flowers blooming? An examination of community energy in the UK. Energy Policy **61**, 977–989 (2013)
N. Simcock, R. Willis, P. Capener, *Cultures of Community Energy. International Case-studies* (British Academy, 2016)
M.M. Sokolowski, Renewable and citizen energy communities in the European Union: how (not) to regulate community energy in national laws and policies. J. Energy Nat. Resour. Law (2020)
T. Sousa, T. Soares, P. Pinson, F. Moret, T. Baroche, E. Sorin, Peer-to-peer and community-based markets: a comprehensive review. Renew. Sust. Energ. Rev. **104**, 367–378 (2019)
B.K. Sovacool, M. Martiskainen, A. Hook, L. Baker, Decarbonization and its discontents: a critical energy justice perspective of four low-carbon transitions. Climate Change **155**, 581–619 (2019)
A. Stevens, C. Macharis, The strategic assessment of driver assistance systems: a multi-criteria approach, in *7th Nectar Conference a New Millennium. Are Things the Same?* (2009)
P.A. Strachan, R. Cowell, G. Ellis, F. Sherry-Brennan, D. Toke, Promoting community renewable energy in a corporate energy world. Sustain. Dev **23**, 96 (2015)
E.O. Taiwo, L. Tozer, Community energy justice: a review of origins, convergence and a research agenda. Energy Res. Social Sci. **123** (2025)
G. Verbong, F. Geels, The ongoing energy transition: lessons from socio-technical, multi-level analysis of the Dutch electricity system (1960-2004). Energy Policy **35**, 1025–1037 (2007)
A.L. Vernay, C. Sebi, F. Arroyo, Energy community business model and their impact on the energy transition: lessons learnt from France. Energy Policy **175** (2023)
J. Vollmer, D. Fouquet. Analysis of legal forms and selection/development of financing and business models for Renewable Energy Communities. (2021). https://energy-cities.eu/wp-content/uploads/2024/07/Report-on-organizational-and-legal-forms-and-business-models-for-RECs.pdf
G. Walker, The role for 'community' in carbon governance. WIREs Clim. Change (2011) https://wires.onlinelibrary.wiley.com/doi/10.1002/wcc.137
G. Walker, P. Devine-Wright, Community renewable energy: what should it mean? Energy Policy **36**(2), 497–500 (2008). https://doi.org/10.1016/j.enpol.2007.10.019
G. Walker, P. Devine-Wright, H. High, B. Evans, Trust and community: exploring the meanings, contexts and dynamics of community renewable energy. Energy Policy **38**, 2655–2663 (2010)
S. Wirth, Communities matter: institutional preconditions for community renewable energy. Energy Policy **70**, 236–246 (2014)
M. Wolsink, The research agenda on social acceptance of distributed generation in smart grids: Renewable as common pool resources. Renew. Sust. Energ. Rev. **16**, 822–835 (2012)

C. Wouters, E.S. Fraga, A.M. James, An energy integrated, multi-microgrid, MILP (mixed-integer linear programming) approach for residential distributed energy system planning- a south Australian case-study. Energy **85**, 30–44 (2015)

M. Yazdani-Damavandi, G. Neyestani, G. Chicco, M. Shafie-Khah, J.P.S. Catalão, Aggregation of distributed energy resources under the concept of multienergy players in local energy systems. IEEE Transactions on Sustainable Energy (2017)

Chapter 3
Key Legal Frameworks and Lessons

3.1 European Norms Operationalize the Energy Communities' Frameworks

Regulations have a pivotal and structural role in the energy transition to frame the complex schemes, conflicts in business model comprehension, usually grid and sharing-related, multiple nature of stakeholders involved, and prosumers' typologies, e.g.., Heffron and Talus (2016). While previous literature focused on technical aspects of self-consumption, mainly on PV prosumers and energy markets, e.g., Butenko (2016), there is still a significant lack in the collective consumption in its regulatory dimensions, e.g., Inês et al. (2020) (see Sect. 2.4).

Until 2018, there had been no specific legal context or references to the instrumentalization and operation of the energy community. Numerous directives and regulations have been issued, aiming to promote energy liberalization and market design for lower electricity prices, starting already three decades ago, e.g., Rotaru (2013), without collective energy undertakings, giving notable relevance to concepts of prosumerism, membership, effective control, flexibility, justice, and citizen participation in decentralized energy systems, e.g., Biresselioglu et al. (2021). In fact, the recent European regulatory analysis gave emphasis on the technical and economic simulations and assessments, acknowledging the factors affecting the innovation in their organizational arrangements but also the non-technical ecosystems and cooperative models to enable the sociocultural shifts, e.g., Sciullo et al. (2022).

Since 1970s, the EU has followed policies on energy efficiency, introducing environmental regulations and concerns for the European community and its members, e.g., Green Paper on Energy Efficiency (2005) (Commission of the European Communities 2005) focused on the efficient use of energy transformation, seeking to act as a catalyst and make a significant contribution to the climate change marathon, by adding energy to the core of European challenges. Toward the same

S. Koutra, *Energy in the Community*, SpringerBriefs in Applied Sciences and Technology, https://doi.org/10.1007/978-3-032-27148-8_3

direction, the European Council established the political targets of the energy and climate packages, precisely in 20% reduction of GHG emissions, 20% renewable use in the energy mixes, and 20% of the energy savings compared to 1990s levels (European Commission 2020).

In 2015, acknowledging the diverse challenges posed by climate change pressures, resource scarcity, and increasing vulnerabilities, the European Commission firstly published in the *Clean Energy Package for all Europeans* (European Commission 2019a), the Fit for 55 Package to reduce GHG emissions by at least 55% by 2030 (compared to 1990s levels) (Erbach and Jensen 2022), and the Repower EU Plan (European Commission 2022), e.g., activating the broader political initiative on a European scale, mobilizing the framework of social and economic capital toward their experimentations and being transposed in local contexts and requirements, introducing new responsibilities to stakeholders, and putting citizens in the core of decision-making processes. This package sets the decarbonization roadmap for 2050 and defines the contribution of each MS by establishing common engagements in local contexts, e.g., the Integrated National Energy and Climate Plan (NECPs) (European Commission 2019b), by promoting public acceptance of renewable energy use and participation in the fight against energy poverty. Nonetheless, already by the 1990s, discussions and scientific debates on energy decentralization had been included in the European agenda with the Directive 96/92/EC (European Parliament 1996), while in 2003, other regulations completed the framework to determine the retail market, e.g., Directive 2003/54/EC (European Parliament 2003a), or the Directive 2003/55/EC (European Parliament 2003b). This framework introduces for the first time the community legislation of the EC concept and acknowledges the term as a way of collective organization, citizen-led, varying on energy activities and forms, including prosumer figures, production, consumption, and storage processes to actively operate the systems and increase efficiency and flexibility in a distributed structure, e.g., López et al. (2024).

Adopted in the early 2000s, different legislative actions (e.g., 2018/2001/EU on the promotion of the RES use (European Commission 2018a), 2019/944/EU on the standard rules for the internal electricity market) were adopted to trace the 2020 roadmap driven by firm commitments (United Nations 2008). In the field of climate and energy, the "20-20-20" strategy set targets for the reduction of GHG emissions, RES sharing and improvement of energy efficiency (European Environment Agency 2021), which brought a set of binding legislation, including the Directive 2009/28/EC on the use of renewable sources in 2009 (European Union 2009), the recast of the Building Energy Performance Directive in 2010 (European Commission 2010a), and the Directive 2012/27/EU on energy efficiency (European Commission 2010b). Along with these, the commitments lead to other legislative acts (2018–2019) establishing particular prerequisites, e.g., Energy Performance of Buildings Directive 2018/884 (European Commission 2018b), the recast Renewable Energy Directive 2018/2001 (known as REDII) (European Commission 2018b), the Energy Efficiency Directive (EU) 2018/2002 (European Union 2018), the Regulation on the Governance of the Energy Union and Climate Action 2018/1999 (European Commission 2018e), etc.

Through these commitments, EU officialized the concept of "energy communities," notably "Citizen Energy Communities (CEC)" and "Renewable Energy Communities (REC)." In reality, two documents form the concepts included in the Internal Electricity Market Directive 2019/944 (European Commission 2019c) and "renewable energy communities" (Renewable Energy Directive, RED) 2018/2001 (European Commission 2018c) aimed to strengthen the role of renewables in support schemes and equal footing with users, which enables for the first time the collective participation in energy systems.

Following RED II, the term "Renewable Energy Community" refers to legal entities composed of members and cooperatives that commonly produce, consume, and manage energy from renewable sources through an "open and voluntary" process, contributing to decarbonization objectives and pursuing environmental and social benefits of self-consumption with non-commercial profits. Precisely, under Article 22/1 of RED II, the right to self-organization of the community is developed through a non-discriminatory process, with specific guidelines that regulate production, consumption, sharing, and selling to ensure justice and maintain the rights of actors as prosumers. Directive 2019/944 (European Parliament 2019) introduced common rules about the generation, transmission, distribution, storage, and electricity supply to create integrated markets for jointly active communities and provide opportunities to secure sustainability on local scales, emphasizing the importance of citizen involvement and regulation of technological processes (Biresselioglu et al. 2021).

Besides, in June 2024, the EU introduced new relevant regulations to improve electricity market operations and design (European Commission 2024b) related to energy sharing to eliminate vulnerabilities and ensure sharing to all users affected by energy poverty and lack of accessibility. In the same year, relevant guidance was published in revised versions of the current regulations to emphasize the active engagement of consumers and enable their participation in decision-making processes in the green transition journey (Nikolic et al. 2025).

These new directives (Table 3.1) promote the roadmap toward clean transitions and the shift of the current energy models to sustainable ones as identified in the Clean Energy for All Europeans Package.

In general terms, EU recognizes the concept as a collective management, human-led, with joint benefits. Although the regulatory framework is broadly harmonized at the European level, national implementations differ considerably across Europe with diverse interpretations on their business models' operations; Western and

Table 3.1 Clean Energy for All European Package ECs framework

Issue	Directive/Regulation	Reference
Energy performance	Directive 2018/844	European Commission (2018c)
Renewable energies	Directive 2018/2001	European Commission (2018b)
Energy efficiency	Directive 2018/2002	European Commission (2018d)
Electricity regulation	Regulation 2019/943	European Commission (2019d)
Electricity Directive	Directive 2019/941	European Commission (2019c)

Southern Europe develop rapid institutional advancements to enable the REC applicability, e.g., Taramboli et al. (2025); examples, such as the case of France, prove the significance of human-centric incentives for the well-established ECs's ecosystems, e.g., Energie Partagée or Centrales Villageoises, e.g., Ranville and Vernayy (2020) with standardized tools, protocols, and mechanisms of democratic governance (Neska and Bielecki 2025), or Italy with clear foundations of technical procedures, and supporting schemes, e.g., Blečić et al. (2025). Conversely, in Central and Eastern Europe, ECs remain scarce in terms of public acceptance and prosumer motivation (Jasiński et al. 2021) but are still not framed by specific policymaking (Jonek-Kowalska 2022), leading to low trust (Zuk and Szulecki 2020), with the consumer perspectives being unclear and embedded in economic initiatives, rather than sociocultural frames.

Another key legislative design of the energy efficiency policies is the provisions of the Ecodesign Regulations (2024/1781) (European Commission 2024c) supporting the energy-related policies to regulate energy efficiency and setting minimum standards for products, machines, and domestic appliances aligned with the EU's energy efficiency rules. Other complementary frameworks and legislative considerations are included in the 2012 EED (European Commission 2012), which merged the Directive 2006/32/EC (European Commission 2006) with the Combined Heat and Power Directive 2004/8/EC (European Commission 2004).

Nevertheless, despite this progress and extensive research on EC operation and beyond, e.g., Vallecha et al. (2021), cross-research studies between technological installations and acceptance suggest the gap in institutional and economic contexts and highlight the importance of contextual factors, as legal or financial stability, where ECs remain marginal and underdeveloped, e.g., Janik et al. (2021). At the same time, and despite the continuous legal provisions, the existing framework *does not elaborate on the ways and exact processes for the energy sharing implementation* with no general rules, entailing uncertainties and multiple risks.

In Europe, a wide range of governmental structures differ in decision-making approaches and liabilities, as well as in their organization, ranging from large cooperatives to collective self-consumption schemes and off-grid island energy systems (Vollmer and Fouquet 2021). Citizen-led initiatives and bottom-up democratic governance lead the vast majority, collectively managed on a voluntary basis to meet local needs and maximize shared benefits. A summarized report on the common legal forms of ECs initiatives is provided in Table 3.2.

The emergence of ECs benefits from the history and experiences of energy cooperatives (Simcock et al. 2016), but also the organization of energy markets, e.g., Lode and Te Boveldt (2022), or cognitive factors, influencing the acceptance of norms, values, and relevant dynamics, e.g., sense of trust, and responsibility, e.g., Walker et al. (2010). Multiple studies have considered the importance of a well-defined and structured legal framework to constitute ECs in an economic, administrative, and technical ecosystem, where the social dimension is usually underrepresented, e.g., Kyriakopoulos (2022).

Table 3.2 Legal forms for energy communities

Legal structure	Characteristics
Energy cooperatives	A fast-growing form of ECs benefits members and is popular in countries where renewables and the energy community are diffuse and traditionally well-established. This is the commonly appearing structure and business model of ECs Most cooperatives are rather small with few members and small-scale energy projects, e.g., PV installations with cases such as Ecopower (Belgium) or Som Energia (Spain) developed on a larger scale with joint investments, as wind turbines for electricity needs. Others, such as the Midcounties cooperative (UK) count more than 500,000 members
Limited partnerships	They allow individuals to share responsibilities and generate profits by participating in the energy community; governance depends on the value of each partner's share (one member does not mean one vote)
Community trusts and foundations	The main aim is to promote social development rather than individual benefits; profits are always used for the community and to foster shared values on a local scale
Housing associations	They offer benefits to tenants in social housing, even if they cannot participate in decision-making. Non-profit associations with social objectives (e.g., alleviation of energy poverty)
Non-profit customer-owned enterprises	They are used by communities that manage independent grid networks. Ideal structures for district heating communities
Public-private partnerships	Local authorities and citizen groups can reach an agreement to ensure energy provision and other benefits for a community
Public utility company	Particularly suited for rural or isolated areas; the public utility company is run by municipalities that manage the utility for taxpayers and citizens

Adapted from Tarpani et al. (2025)

An overview of standard types and cases of European organizational forms and business models developed, including institutional settings, is detected in countries, such as Belgium, and cooperative business models, established by private and public contributions. Germany is among the pioneers of collective actions, with a long history of collaborative operations, counting more than 200,000 members, and has developed self-consumption models, primarily at the building scale. In the Netherlands, the most prevalent forms of collective energy actions have been developed through energy cooperatives or associations since 2015, focusing primarily on renewables, with more than 600 cases today. A unique case in the analysis is still Norway, due to its high renewable share in the electricity mix, with more than 30 local community projects with limited citizen engagement, nonetheless. Another particular case is Spain, with a long tradition of energy cooperatives since the 1920s to advance rural electrification. In other countries, such as Italy, they are still in the early stages and are establishing thriving first communities through bottom-up models and traditional cooperatives (Vollmer and Fouquet 2021).

3.2 Legal Frameworks on ECs at National Levels

When reflecting on information across national levels in European Member States, very few studies are acknowledged regarding the transposition of European commitments. Multiple complexities are identified in this level, deriving from the difficulties in definitions on REC and CEC concepts, but also across sectors of energy efficiency and renovation, and supporting governance schemes, which necessitate integrated regulations upon the removal of inequalities and the promotion of justice, awareness, capacity building, and inclusiveness (European Commission 2024a).

So far, the national transposition of EU Directives on EC topics has a twofold challenge:

- Ensure compliance with the national and governance requirements responding to local needs.
- Enable ECs to conduct their activities, including sharing with limited research on concrete implementation of this, e.g., Sokolowski (2020).

Although the transposition of the abovementioned engagements is, respectively, developed at varying stages between the Member States, Nordic countries accelerate the process in their national frameworks with the introduction of renewables' promotion (e.g., Denmark, 2021), and the integration of specific contexts, as the case of Local Energy Community (LEC) in Finland (RESCOOP.EU 2022). Conversely, countries such as Sweden and Norway have not officially recognized the context of ECs to date, primarily due to a robust, long-term history of energy system price management; hence, few EC projects are at an early implementation stage. Similar is the situation in Sweden, where the contextualized definitions of ECs are in an infancy phase as well and have not been included in the official regulatory frameworks yet; nonetheless, some ECs projects are already active with a growing public interest (Nordic Energy Research 2023).

In other countries of Europe, outside the Nordics, the transposition varies as well, with Austria to be most far ahead and the establishment of the Renewable Energy Act (Federal Ministry of Economy, Family and Youth 2010), being a pioneer in new market roles with more than 4000 established by 2025 developed in three types: (1) Joint communities with common production and electricity use on same ownership and infrastructure, (2) local energy communities, involving as well public authorities and private companies, and (3) citizen energy communities; most of them organized as cooperatives or associations and basically lie on solar energy potential and less on wind and biomass (Berggren Eriksson et al. 2023). In Sweden, the main legislative directive is the Electricity Act (Swedish Energy Markets Inspectorate 2022), which promotes energy efficiency measures and renewable energy frameworks. Relevant regulations are found in Germany; examples are the Renewable Energy Sources Act (Bundeswirtschafts Ministerium 2017) or the Energy Industry Act (Bundesministerium der Justiz und fur Verbraucherschutz 2005) and laws on cooperatives to enable ECs operation.

References

S. Berggren Eriksson, T. Witt, E. Van der Linden, L. Saes, L.E. Arvefjord, D. Heckenberg, T. Iglauer, L. Sutinen, E. Hanning, G. Melin, Energy communities. Nordic Energy Res. (2023) https://pub.norden.org/nordicenergyresearch2023-03/nordicenergyresearch2023-03.pdf

M.E. Biresselioglu, S.A. Limoncuoglu, M.H. Demir, J. Reichl, K. Burgstaller, A. Sciullo, Legal provision and market conditions for Energy Communities in Austria, Germany, Greece, Italy, Spain and Turkey: a comparative assessment. Sustainability **13** (2021)

I. Blečić, A.S. Carrus, E. Congiu, G. Desogus, E. Muroni, V. Saiu, Renewable energy communities design: a decision support tool for integrated impact assessment. Insights from the first REC in Cagliari, Italy. J. Clean. Prod **510**, 145600 (2025)

Bundesministerium der Justiz und fur Verbraucherschutz. (2005). Act on electricity and gas supply (Energy Industry Act—EnWG). https://www.gesetze-im-internet.de/enwg_2005/BJNR197010005.html

Bundeswirtschafts Ministerium. (2017). Renewable Energy Sources Act (EEG 2017). https://www.bundeswirtschaftsministerium.de/Redaktion/EN/Downloads/renewable-energy-sources-act-2017.pdf?__blob=publicationFile&v=1

A. Butenko (2016). *User-centered innovation and regulation framework: Energy prosumers' market access in EU regulation* [Discussion paper]. https://papers.ssrn.com/sol3/papers.cfm?abstract_id=2797545

Commission of the European Communities. *Green Paper on Energy Efficiency or Doing More with Less*. (2005). https://www.ab.gov.tr/files/ardb/evt/1_avrupa_birligi/1_6_raporlar/1_2_green_papers/com2005_green_paper_on_energy_efficiency.pdf

Erbach, G., & Jensen, L. *Briefing towards climate neutrality. Fit for 55 package*. (2022). https://www.europarl.europa.eu/RegData/etudes/BRIE/2022/733513/EPRS_BRI(2022)733513_EN.pdf

European Commission. *Directive 2004/8/EC of the European Parliament and of the council of 11 February 2004 on the promotion of cogeneration based on a useful heat demand in the internal energy market and amending directive 92/42/EEC*. (2004). https://eur-lex.europa.eu/eli/dir/2004/8/oj/eng

European Commission. *Directive 2006/32/EC of the European Parliament and of the Council of 5 April 2006 on energy end-use efficiency and energy services and repealing Council Directive 93/76/EEC*. (2006). https://eur-lex.europa.eu/eli/dir/2006/32/2014-06-05/eng

European Commission. *Directive 2010/31/EU of the European Parliament and of the Council of 19 May 2010 on the Energy Performance of Buildings*. (2010a)

European Commission. *Directive 2012/27/EU of the European Parliament and of the Council of 25 October on Energy Efficiency*. (2010b)

European Commission. *Directive 2012/27/EU of the European Parliament and of the Council of 25 October 2012 on energy efficiency, amending Directives 2009/125/EC and 2010/30/EC and repealing Directives 2004/8/EC and 2006/32/EC Text*. (2012). https://eur-lex.europa.eu/eli/dir/2012/27/oj/eng

European Commission. *Directive (EU) 2018/844 of the European Parliament and of the Council of 30 May 2018 Amending Directive 2010/31/EU on the Energy Performance of Buildings and Directive 2012/27/EU on Energy Efficiency* [Directive]. (2018a)

European Commission. *Directive (EU) 2018/2001 of the European Parliament and of the Council of 11 December 2018 on the Promotion of the Use of Energy from Renewable Sources*. (2018b)

European Commission. *Directive (EU) 2018/2001 of the European Parliament and of the Council of 11 December 2018 on the promotion of the use of energy from renewable sources*. (2018c). https://eur-lex.europa.eu/legal-content/EN/TXT/?uri=CELEX%3A32018L2001

European Commission. *Directive (EU) 2018/2002 of the European Parliament and of the Council of 11 December 2018 amending Directive 2012/27/EU on energy efficiency*. (2018d). https://eur-lex.europa.eu/eli/dir/2018/2002/oj/eng

European Commission. *Regulation (EU) 2018/1999 of the European Parliament and of the Council of 11 Dec 2018 on the Governance of the Energy Union and Climate Action*. (2018e)

European Commission. *Clean energy for all Europeans package.* (2019a). https://wayback.archive-it.org/12090/20241209144917/https://energy.ec.europa.eu/topics/energy-strategy/clean-energy-all-europeans-package_en

European Commission. *National Energy and climate plans.* (2019b). https://commission.europa.eu/energy-climate-change-environment/implementation-eu-countries/energy-and-climate-governance-and-reporting/national-energy-and-climate-plans_en

European Commission. *Regulation (EU) 2019/941 of the European Parliament and of the Council of 5 June 2019 on risk-preparedness in the electricity sector and repealing Directive 2005/89/EC.* (2019c). https://eur-lex.europa.eu/eli/reg/2019/941/oj/eng

European Commission. *Regulation (EU) 2019/943 of the European Parliament and of the Council of 5 June 2019 on the internal market for electricity (recast).* (2019d). https://eur-lex.europa.eu/eli/reg/2019/943/oj/eng

European Commission. *Europe 2020. A strategy for smart, sustainable and inclusive growth.* (2020). https://ec.europa.eu/eu2020/pdf/COMPLET%20EN%20BARROSO%20%20%20007%20-%20Europe%202020%20-%20EN%20version.pdf

European Commission. *Communication from the Commission to the European Parliament, the European Council, the Council, the European Economic and Social Committee and the Committee of the Regions. REPowerEU Plan.* (2022). https://eur-lex.europa.eu/legal-content/EN/TXT/HTML/?uri=CELEX:52022DC0230

European Commission. *A roadmap to developing a policy and legal framework that enables the development of Energy Communities.* (2024a). https://build-up.ec.europa.eu/system/files/2024-03/c6990021-e30c-4b29-9dd9-147989a1869b_en.pdf

European Commission. *Directive (EU) 2024/1711 of the European Parliament and of the Council of 13 June 2024 amending Directives (EU) 2018/2001 and (EU) 2019/944 as regards improving the Union's electricity market design.* (2024b). https://eur-lex.europa.eu/eli/dir/2024/1711/oj/eng

European Commission. *Regulation (EU) 2024/1781 of the European Parliament and of the Council of 13 June 2024 establishing Directive (EU) 2020/1828 and Regulation (EU) 2023/1542 and repealing Directive 2009/125/EC.* (2024c). https://eur-lex.europa.eu/eli/reg/2024/1781/oj/eng

European Environment Agency. *EU achieves 20-20-20 climate targets, 55% emissions cut by 2030 reachable with more efforts and policies.* (2021). https://www.eea.europa.eu/highlights/eu-achieves-20-20-20

European Parliament. *Directive 96/92/EC of the European Parliament and of the Council of 19 December 1996 concerning common rules for the internal market in electricity.* (1996). https://eur-lex.europa.eu/eli/dir/1996/92/oj/eng

European Parliament. *Directive 2003/54/EC of the European Parliament and of the Council of 26 June 2003 concerning common rules for the internal market in electricity and repealing Directive 96/92/EC-Statements made with regard to decommissioning and waste management activities.* (2003a). https://eur-lex.europa.eu/eli/dir/2003/54/oj/eng

European Parliament. *Directive 2003/55/EC of the European Parliament and of the Council of 26 June 2003 concerning common rules for the internal market in natural gas and repealing Directive 98/30/EC.* (2003b). https://eur-lex.europa.eu/eli/dir/2003/55/oj/eng

European Parliament. *Directive (EU) 2019/944 of the European Parliament and of the Council of 5 June 2019 on common rules for the internal market for electricity and amending Directive 2012/27/EU (recast).* (2019). https://eur-lex.europa.eu/legal-content/EN/TXT/?uri=CELEX:32019L0944

European Union. *Directive 2009/28/EC of the European Parliament and of the Council of 23 April 2009 on the Promotion of the Use of Energy from Renewable Sources and Amending and Subsequently Repealing Directives 2001/77/EC and 2003/30/EC.* (2009)

European Union. *Directive (EU) 2018/2002 of the European Parliament and of the Council of 11 December 2018 Amending Directive 2012/27/EU on Energy Efficiency.* (2018)

Federal Ministry of Economy, Family and Youth. *National Renewable Energy Action Plan 2010 for Austria under Directive 2009/28/EC of the European Parliament and of the Council. National*

Renewable Energy Action Plan 2010 for Austria. (2010). https://build-up.ec.europa.eu/sites/default/files/content/national_renewable_energy_action_plan_austria_en.pdf

R.J. Heffron, K. Talus, The evolution of energy law and energy jurisprudence: Insights for energy analysts and researchers. Energy Res. Soc. Sci. **19**, 1–10 (2016)

C. Inês, P.L. Guilherme, M.G. Esther, G. Swantje, H. Stephen, H. Lars, Regulatory challenges and opportunities for collective renewable energy prosumers in the EU. Energy Policy **138**, 111212 (2020)

A. Janik, A. Ryszko, M. Szafraniec, Determinants of the EU citizens' attitudes towards the European energy Union priorities. Energies **14** (2021)

J. Jasiński, M. Kozakiewicz, M. Soltysik, Determinants of energy cooperatives' development in rural areas-evidence from Poland. Energies **14** (2021)

I. Jonek-Kowalska, Multi-criteria evaluation of the effectiveness of energy policy in Central and Eastern European countries in a long-term perspective. Energ. Strat. Rev. **44**, 100973 (2022)

G. Kyriakopoulos, Energy communities overview: managerial policies, economic aspects, technologies, and models. J. Risk Financ. Manag. **15** (2022)

M.L. Lode, T. Te Boveldt, Transition perspective on energy communities: a systematic literature review and research agenda. Renew. Sust. Energ. Rev., 1–12 (2022)

I. López, N. Gotia-Zabaleta, A. Milo, J. Gómez-Cornejo, I. Aranzabal, H. Gaztañaga, E. Fernandez, European energy communities: characteristics, trends, business models and legal framework. Renew. Sust. Energ. Rev., 197 (2024)

E. Neska, M. Bielecki, Bridging the implementation gap: Expert, systemic, and policy perspectives on renewable energy communities in Poland. Energ. Strat. Rev. **62** (2025)

J. Nikolic, O. Clemmensen, M. Gerritsen, K. Sperling, H. J. Kooij. *COPPER project: energy communities: Inventory of barriers and opportunities and best practice.* (2025). https://dutpartnership.eu/sites/default/files/2025-11/COPPER%20ENERGY%20COMMUNITIES%20INVENTORY%20OF%20BARRIERS%20OPPORTUNITIES%20AND%20BEST%20PRACTICE.pdf

Nordic Energy Research. (2023). Energy communities. https://pub.norden.org/nordicenergyresearch2023-03/netherlands.html

A. Ranville, A.L. Vernayy, *(2020). Research Report on 'Renewable Energy Cooperative' in France* (SONNET, EU Horizon, 2020). https://shs.hal.science/halshs-03356278/document

RESCOOP.EU. *Finland. REC & CEC Definitions. Enabling Frameworks & Support Schemes.* (2022). https://www.rescoop.eu/policy/transposition-tracker/rec-cec-definitions/finland

D.V. Rotaru, *A Glance at the European Energy Market Liberalization* (CES Working Papers, Centre for European Studies, 2013) https://ideas.repec.org/a/jes/wpaper/y2013v5i1p100-110.html

A. Sciullo, G.W. Gilcrease, M. Perugini, D. Padovan, B. Curli, J.S. Gregg, O. Arrobbio, E. Meynaerts, S. Delvaux, L. Polo-Alvarez, Exploring institutional and socio-economic settings for the development of energy communities in Europe. Energies **15** (2022)

N. Simcock, R. Willis, P. Capener, *Cultures of Community Energy. International Case-studies* (British Academy, 2016)

M.M. Sokolowski, Renewable and citizen energy communities in the European Union: how (not) to regulate community energy in national laws and policies. J. Energy Nat. Resour. Law (2020)

Swedish Energy Markets Inspectorate. *Sweden's electricity and natural gas market.* (2022). https://ei.se/download/18.31b721ca18388fe22ad2d3f/1664799315977/Sweden%27s-electricity--and-natural-gas-market-2021-Ei-R2022-07.pdf

G. Taramboli, L. Campagna, C. Bergonzi, F. Bovera, V. Trovato, M. Merlo, G. Rancilio, Renewable Energy Communities: frameworks and implementation of regulatory, technical, and social aspects across EU member states. Sustainability (2025)

E. Tarpani, T. Hong, I. Pigliautile, W. Zhang, K. Sun, B. Pioppi, A.L. Pisello, On renewable energy community implementation in historic cities: a city-scale validated model. *Energy Build.* **338**. Scopus (2025). https://doi.org/10.1016/j.enbuild.2025.115709

United Nations, *Kyoto Protocol Reference Manual on Accounting of Emissions and Assigned Amount* (2008). https://doi.org/10.5213/jkcs.1998.2.2.62

H. Vallecha, D. Bhattacharjee, J. Osiri, P. Bhola, Evaluation of barriers and enablers through integrative multicriteria decision mapping: developing sustainable community energy in Indian context. Renew. Sust. Energ. Rev. **138** (2021)

J. Vollmer, D. Fouquet. Analysis of legal forms and selection/development of financing and business models for Renewable Energy Communities. (2021). https://energy-cities.eu/wp-content/uploads/2024/07/Report-on-organizational-and-legal-forms-and-business-models-for-RECs.pdf

G. Walker, P. Devine-Wright, H. High, B. Evans, Trust and community: exploring the meanings, contexts and dynamics of community renewable energy. Energy Policy **38**, 2655–2663 (2010)

P. Zuk, K. Szulecki, Unpacking the right-populist threat to climate action: Poland's pro-governmental media on energy transition and climate change. Energy Research & Social Science (2020)

Chapter 4
Portfolio and Analysis of European Notable Examples

Reporting more than 8000 projects, ECs' growth does not seem homogeneous across Europe, with more cases appearing in Germany and the Netherlands, followed by Austria, Switzerland, Denmark, France, Great Britain, Poland, Spain, and Sweden (Padovan et al., 2019); hence, ECs' number in Europe remains approximate, e.g., Bauwens (2017). Caramizaru and Uihlein (Caramizaru & Uihlein, 2020) presented 24 cases from nine (9) European countries, considering broader targets, boundaries, and scales, ranging from large to small citizen-organized energy cooperatives. While ECs have already developed in different countries moving toward a community-led business model for a long time in Europe (see Sect. 2.1), basically in the form of cooperatives, e.g., Belgium, Denmark, Germany, the Netherlands, Spain, Sweden, and elsewhere, there is still lack of supportive mechanisms (legislative and societal) to others with the majority of the EU Members States to have little or even no differentiation made between REC and CEC in their general and practical discourses, usually referred simply as "energy communities," but with significant convergencies in scale, size, perimeter, and format.

Considering the importance of cooperative structures, as the common EC configuration, countries, such as Germany, the Netherlands, or the United Kingdom, have witnessed a significant amount of EC markets integrating prosumers in decision-making developments (Hendricks & Mesquita, 2019). Literature review enabled the analysis of different ECs' patterns, using taxonomy and typologies, as, for instance, Weber (Weber, 2011) proposed:

- Level of business model maturity and operation, e.g., Rossetto et al. (2022)
- Type of energy activities (e.g., generation, sharing, distribution, etc.)
- Geographical location or purpose, e.g., Moroni et al. (2019)

North-Western Europe is a pioneer in community initiation with powerful national policies, citizen-oriented to promote decentralized systems, e.g., Bauwens et al. (2016), enabled as well by the long-lasting tradition of cooperative operations for energy management issues, e.g., Reis et al. (2021). Around 3000 organizations

S. Koutra, *Energy in the Community*, SpringerBriefs in Applied Sciences and Technology, https://doi.org/10.1007/978-3-032-27148-8_4

have been reported in European projects (Huybrechts & Mertens, 2014), mostly matured in Western Europe, e.g., Denmark, the Netherlands, Germany, Italy, and the UK, and across models of fossil energy, as in many cases in Eastern and Central Europe (Becker & Kunze, 2014). Evidence-based studies, e.g., Berka and Creamer (2018) and Lode, Coosemans et al. (2022), highlighted the institutional and cultural relatedness as a key indicator for the EC development, pushing broader drivers for shared benefits, while other scholars, e.g., Bouzarovski and Simcock (2017) and Garvey et al. (2022), reflect the importance of spatial and geographical contexts related to the resource distribution as motivation for EC development, or even economic and social circumstances related to energy poverty, e.g., Cutter et al. (2003).

As one of the leading countries on the EC projects, Denmark is an important European energy hub and has a decentralized governance system with a liberalized electricity market. At the same angle, Germany established ambitious targets and adopted specific measures to promote the energy transition (in 2016, there were more than 1000 registered energy—mainly wind power-based—initiatives in Germany), thanks to a well-established, alternative, and autonomous energy movement (Gancheva et al., 2018). In fact, in the EC gray and scientific documentation, much attention has been given to these countries, e.g., Germany, e.g., Holstenkamp and Kahla (2016), Denmark, e.g., Mey and Diesendorf (2018), or United Kingdom, e.g., Berka (2018), with fewer cases appearing for French cases; exceptions are the studies by Sebi and Vernay (Sebi & Vernay, 2020) in 2020 and the comparative study with the Netherlands case (Vernay & Sebi, 2020).

Simcock et al. (Simcock et al., 2016), in their reports, introduced cases and lessons learned, acknowledged as leaders in the community-level energy sharing and management of international cases, e.g., Middelgrunden, as the largest community-led wind farm (Denmark) in 1997 with a joint investment of 10.2 MW (REScoop, 2016), and in Germany (EWS) as the first civil-society initiative for decentralized systems in local grids (EWS, 2016), pioneering showcasing on how regulatory frameworks facilitate the EC operation, comprised of 20 wind turbines, which delivers about 4% of the power for the city of Copenhagen (Laurel, 2024), and Hvide Sande Wind Farm (Denmark), or the Bioenergy Village Jühnde (Germany), Ecopower (Belgium), Brixton Energy (UK), or elsewhere in the world, e.g., Buan County (Korea). Strong policies supporting liberalization of energy markets, decentralization of energy systems, and funding mechanisms were crucial drivers for ECs' development.

Among the most acknowledged RES cooperatives in Europe, the following are cited: Enercoop (France) (Enercoop, 2025) (more than 40,000 members), Ecopower (Belgium) (Ecopower, 2025) (more than 50,000 members), Energy4All (UK) (supports 30 ECs' projects) (Energ4All, 2025), etc., heavily influenced by normative frameworks established along with relevant institutional supports to enable technical strategies and embrace technological innovation collectively based.

Several typologies, e.g., Hewitt et al. (2019), stemming from academic documentation define solar, wind, hydro, or e-mobility communities, or about the geographic context (rural or urban), size (number of members), or purposes and activities (generation, consumption, etc.) or the legal forms (cooperatives, associations, non-profit organizations, etc.) (Koltunov et al., 2023), clusters divided into renewables and non-renewables (Braunholtz-Speight et al., 2020), or even other taxonomies categorized by physical location in place and non-place based, as proposed by Moroni et al. (2019c). Reis et al. (2021b) proposed eight (8) models: cooperatives, prosumerism, local markets, collective generation, third-party, flexibility aggregation, companies, and e-mobility structures. Other studies organize the ECs by (a) the type of business model, e.g., Rossetto et al. (2022), as nodes of polycentric systems (Gui et al., 2017), (b) activities, e.g., Moroni et al. (2019c), (c) geographical location (Petersen, 2016), (d) small-scale enterprises characterized by collectivity, e.g., Romero-Rubio and de Andrés Díaz (2015); whatever is the definition, the key dimensions of the term consist systemic approaches of the beneficiaries and end users, where the role of private sectors should be clearly configured, e.g., Van der Horst (2008) (Table 4.1). Reviewing the European cases, most of them are grouped under the Re Co-Op Federation, which consists of more than 1500 cooperatives with a long-lasting history already in Northern Europe, especially in activities of self-consumption and generation, e.g., Wierling et al. (2018).

Studies, e.g., Koltunov et al. (2023), prove the heterogeneity of ECs' configurations in Europe, highlighting the importance of policy frameworks and citizen participation levers. Cooperatives in Belgium and Spain are powerful in fighting energy poverty and promoting sustainable development. Catalyst for the rapid spread of cases in Italy and Spain has been the transposition of ECs' Directives to clarified national frames, the active stakeholder involvement, and alternative business models, including public sector "player" as municipalities and non-governmental organizations, or even research clusters (case of Italy). To this, government-supported actions to promote market efficiency evolved business models and increased energy productivity, achieving relevant societal and environmental outcomes profitable for the "common good." Agrawal and Hockerts (Agrawal & Hockerts, 2021) assessed the investment attractiveness and the adaptability of existing models to social innovation processes that foster collective dynamics of profitability.

Stepping stone for the EC development has been the European Federation of Energy Cooperatives (REScoop.eu), as the organization body, comprising national federations and promoting activities of energy sharing to meet local needs, represented mainly by national associations, as REScoop (Belgium), Energie Samen (the Netherlands), DGRV (Germany), among others (Koltunov et al., 2023). Cases, as the Enostra (Italy) and Ecopower (Belgium) are documented as the largest initiatives (in their respective countries) with interesting similarities in the organizational models acting as energy suppliers and based on mutuality principles and social values with democratic systems and inclusive design targeting the people empowerment to energy and sustainability concerns (Dudka et al., 2024) and concentrated on priorities of social challenges.

Table 4.1 Typical examples of European energy communities

Country	Case study	Background and motivation	Results
Spain (Montojo, 2020)	Crevillent community	Solar PVs, objective for 100% clean energy production and distribution	Supplies for more than 14,000 prosumers, 13,4 MW of solar power, avoiding more than 35,000 t of CO_2 annually
Italy (Legambiente, 2022)	Lodi	Solar-powered community committed to decarbonization and promotion of local production	Integration of storage to maximize energy efficiency
Denmark	Middelgrunden wind farm (offshore wind farm 3 km from Copenhagen harbor)	Development of a wind turbine initiative and a partnership with local utilities	20 turbines of 2 MW each providing 40 megawatts of electricity (equivalent to 3% of the capital's electricity needs)
	Hvide Sande wind farm (small port village of 3000 people and fifth largest port)	Different from other energy developments Decentralized planning systems democratically developed	3 MW wind turbines installed on a beach Local heat and power plant and district heating system 65% of all households have district heating Electricity sold to the national grid
	Marstal Fjernvarme (renewable energy island)	Old fossil fuel-based heating systems replaced by renewables Thermal storage to conserve solar heat throughout the winter	Heat provider from 100% renewable energy sources
Germany	Bioenergy Village Junhnde (first in Germany in 2000)	Bioenergy facility owned locally and collectively by residents Commitment by local actors and involvement	Local bioenergy plant to supply heat and power to residents 700 kW CHP generator for the electricity production supplied to the public grid 550 kW woodchip boiler for heating around the local district network Production of plants to 70% of the heating demand 60% CO2 reductions

(continued)

Table 4.1 (continued)

Country	Case study	Background and motivation	Results
United Kingdom	Wiltshire wildlife	Motivation for biodiversity protection (Barlow, 2024)	Two ground-mounted solar PV farms of 1 MW and 9.1 MW
	Horshader community wind turbine (Scotland)	Community not-for-profit organization aiming at supporting local development and regeneration First community-owned turbine in the Western isles	Single 900 kW wind turbine
Belgium (Flanders)	Ecopower	Different schemes for multiple energy projects (electricity supplies)	Energy generation from hydro, solar, and wind supplies electricity Existing projects include 20 wind turbines, 320 solar PV installations in schools and homes, and a cogeneration plant
Estonia	Energy efficiency renovation (Talin)	Renovation energy efficiency project and small-scale energy production	Solar PV installations Insulation of façades and roofs, triple-glazed windows, 15 kW PV system (roofs)
	First solar community (Kagu)	Project in the initial phase (intention for a solar PV farm model) (CO_2 community, 2018)	Project economically viable with a long payback period
Finland	Alapua village	Investment in a CHP plant for energy production (generation of 200MWh of heat and 70,000–80,000kWh of electricity) (CO_2 Community, 2018)	Installation of a CHP plant for heating and electricity for educational buildings
Lithuania	Wind community project (Smalininkai village)	Wind power initiatives and developments with ambitions for wind power capacity	First community wind project owned by a local community Power plant consisted of a 250 kW wind turbine with technical failures

Adapted from Simcock et al. (2016)

4.1 Denmark and Its Long History of Energy Cooperatives

Denmark has one of the longest histories among European Member States in the operation of energy cooperatives and the share of energy investments, with more than 600 cases organized by citizens, projects of heating and electricity production supported by citizen ownership models for wind turbines (more than 10) and district heating systems (Gorroño-Albizu et al., 2019). In the study of 2019 published by SWECO (Sweco, 2019), Denmark was estimated as the country with the highest distributed electricity production for self-consumption in 2017 among the Nordic countries, with the majority stemming from wind power. This combination of successful energy governance with the high amount of energy production enables the EC acceleration, resulting in more than 600 EC projects (Wierling et al., 2023). In terms of EC models, Denmark promotes the collective grid assets, while the EC is often organized as an association, partnership, cooperative, or company, and initiated through already established groups of people, as municipalities, housing cooperatives, or eco-villages, considered as a communal institution with collective ownership and overarching motivations for self-sufficiency beyond the technical issue related to consumption and production activities with particular focus on heating and electricity sharing. Another interesting governance case in Denmark is the housing cooperative, framed with specific regulations for PV and heat pump installations (Nordic Energy Research, 2023).

Practical barriers mentioned by scholars for the case of Denmark are basically related to insufficient opportunities for cost-effective electricity sharing and autonomy by ECs at their distribution networks, with the creation of internal (micro)grid, and limited technical knowledge to deal with, especially regarding the relevant data collection and context-specific business models (Nordic Energy Research, 2023).

Typical examples of EC initiatives in Denmark are:

- *Case 1: Karise Permatopia (south of Copenhagen)* (Flexalen, 2018; Nordic Energy Research, 2023): Developed as an eco-village with a common geothermal heating system based on decentralized local production (basically geothermal heat) exclusively from renewable resources, established in 2018 with the ambition of circular energy systems developments.
- *Case 2: Avedore* (Nordic Energy Research, 2023): *First citizen energy community* in Denmark (south of Copenhagen), founded in 2020 by local actors, as part of the development plan called "Avedore Green City" supported by the ELENA project related to the district heating network renovation in social housing districts of the site and the increase in energy efficiency levels and relevant measures, e.g., improvements on building envelopes, insulation, among others (Christiansen, 2023).

4.2 The Netherlands and the Model of Cooperatives

In the Netherlands, no single definition of the ECs' conceptualization is found, with most cases fitting the definition of RECs proposed in EU regulations, and recognized by their role as a catalyst for the energy transition of the country, but with still unexplored uncertainties and practical barriers in their setting-up, in particular in terms of legislative unclarities.

The first efforts to establish local energy communities in the Netherlands appeared in the late 1980s, often involving wind turbines. The liberalization of the energy market in 2004 increased local initiatives and accelerated processes, with the number of ECs reaching 25 recently (Kooij et al., 2018). ECs are organized in diverse ways as well in the Netherlands, with the main forms being the cooperatives, associations, and sometimes companies, encouraged by the national regulations, e.g., the new Electricity Act Experiments Scheme (Climate Change Laws of the World, 2018), while in 2023, the Netherlands counted 714 ECs in cooperative structures, 146 solar and 105 wind projects (Nikolic et al., 2025); however, several barriers are observed in practicalities, including the challenges on financial schemes, lack of technical and organizational knowledge, but also complexities on business models' interpretations.

Pioneer ECs cases in the Netherlands are (Nordic Energy Research, 2023):

- *Case 1: Republica Papaverweg*: Composed of a variety of building typologies with a smart grid sharing and integration of circularity principles.
- *Case 2: Schoonschip (North Amsterdam)*: Example of a self-sufficient community of 46 water houses and common use of a smart grid.

4.3 France: Business Models of ECs and the Case of EnerCoop

The French government has formally acknowledged the major contribution of energy communities to energy decentralization by transposing the European Directives into law no. 2019–1147 on energy and climate, which addresses climate targets through bottom-up approaches and introduces citizen participation in the transition process. Between 2016 and 2019, ECs in France (under the managerial umbrella of Energie Partagée) doubled their production and increased their capital by over 50% and numbers from 4820 to 5947 (Energie Partagée, 2020).

Vernay et al. (Vernay et al., 2023) reported five (5) business models of energy communities based on the diversity of their activities, the nature of forms and types of stakeholders, the geographical scope, and the types of members involved:

- **Local Integrated Energy Suppliers**: Through legal entities, where public authorities have at least a controlling stake in consumers and maintain control of decision-making processes; the communities are developed within the city's or inter-municipality levels' boundaries.

- **Collective Self-Consumption**: Principally developed by local governments, social housing associations, and organized production and consumption activities targeting autonomy at the local level.
- **Neighboring Energy Supply**: Project developers and energy suppliers enable the local generation from close renewable sources as a service for local consumers and differentiates itself from other forms of ECs business models.
- **Citizen Energy Production**: CECs members in collaboration with local public authorities for increasing the capacity of renewable energy uses.
- **Cooperative Energy Supply**: Energy suppliers in diverse entities, such as NGOs (e.g., Enercoop), independent energy producers, or CECs (e.g., Enargia or Energie de Nantes), enabling the consumption of renewable electricity.

EnerCoop is France's largest Renewable Energy Cooperative, started operations in 2005, and remains active and continues to grow. It is known as the country's largest renewable energy supplier, founded by 55,000 members and owners representing different stakeholders of the energy system, including individual natural persons and business organizations. It focuses on several technologies (e.g., 21% in solar PV, 21% in hydropower, 58% in wind, and 0.1% in biomass) (*Coopératives d'énergie renouvelable, citoyenne et solidaire*, 2020). The cooperative has reached 355 MW power capacity through purchasing renewable electricity from 470 local energy producers. In 2021, EnerCoop supplied electricity to 101,000 customers. EnerCoop is registered as a cooperative society of collective interest (SCIC), which is a legal status particular to France and has a democratic and ownership structure, analogous to that of a typical Energy Cooperative.

4.4 Other Interesting Cases in Europe

Another notable case to consider is the Netherlands with two local citizen energy cooperatives, in Zeeuwind (Zeeuwind coop, 2021) (more than 2500 members) and Deltawind (Deltawind, 2021) (more than 3000 members), with no single definition or ECs' operational model. Being appeared already in 1980s, the first local communities in the Netherlands emerged with wind turbine projects, while in 2010s, the integration of innovative funding schemes, such as the SDE+ subsidy, enabled their replicability in diverse forms of organization, as foundations, associations of owners, with the frequent involvement of private sector in the governance structures. Currently, many efforts for the alleviation of legislative complexities are being developed (e.g., introduction of Energy Law for encompassing all energy carriers, etc.) to enable the sharing activities, but with still unclear processes and critical lacks. Other barriers recorded in the Netherlands for EC scalability concern the financial challenges to secure upfront investments and private involvement, but also complicated issues of knowledge transferability and learning developments toward stakeholders' activation make the EC development still difficult. Pioneering cases in the Netherlands are the examples of Republica Papaverweg (smart grid share),

Schoonschip (water houses and use of smart grid targeting autonomy), and Earth Houses (communal areas) (Nordic Energy Research, 2023).

In Spain, the EC is defined as a legal entity established in the legal system with the first applications of renewable communities less than ten years ago (Primer Modelo de Autoconsumo Comunitario—Red Eléctrica de España, 2021), excluding the historical cooperatives of Valencia, and currently has more than 100,000 members within its ten (10) distinct autonomous regions counting over 180 cases, many in the planning stages, or others in place with already ongoing activities of generating and sharing energy, using basically as vectors the photovoltaic, hydraulic, or biomass energy (Delicado et al., 2017).

ECs in Spain aim to promote decentralized systems, citizen-led, and encourage the shared self-consumption and governance. Others, such as the case of Rupiá (Catalonia), Albalat dels Sorells (Valencia), and Río Monachil (Andalucia), focus their actions on the fight against energy poverty by involving vulnerable households in their structures, with significant differences in their activities, participation levels, and ownership status; for instance, in the case of Rupiá the city council is the only owner of the installation and determines the distribution among prosumers, in the case of Río Monachil the PV installations are managed by public buildings (without being part of the members) with absence of participation by the city council and costs sharing. Spain counts as well rural cases, such as the Mediona, Bufalvent Industrial Estate, and Vilanoveta (Parreño-Rodriguez et al., 2023).

An interesting case of Spain is the Alumbra energy community (located in Arroyomolinos de Leon, province of Huelva) with a declining population and numerous vulnerabilities developed by local residents to promote collective self-consumption and set up photovoltaic technologies through crowdfunding schemes with outstanding outcomes of a functional democratic model bringing out cost and energy savings, coupled with learning and educational programs for best practice sharing.

The case of EC Torreblanca Illumina is another remarkable case developed in the southern zone of Seville to address the energy poverty challenge as well, considering that the Torreblanca area has one of the lowest average incomes in Spain. The community was established in 2020 by the association "Asociación Torreblanca Ilumina, Comunidad Ciudadana de Energía y Aprendizaje" as a learning and community-building project to raise awareness of energy conservation and environmental risks between the school and the district surroundings. Publicly and privately funded, it received public interest and was developed as a pilot demonstrator, involving two photovoltaic solar systems with 70 panels to provide free energy for 13 households and save more than 40% on their electricity bills (Delicado et al., 2017).

In the same lens, Portugal develops the first communities in 2019 with the transposition of the EC Directives in its national framework (Decree Law 162/2019 (Decreee Law No 162/2019, 2019)) under the format of collective self-consumption and REC for the establishment of collective legal entities in a bloc of apartments or housing groups in urban or rural areas in proximity to the generation site. The progress on regulatory frameworks published in the Decree Law 15/2022 (Decree Law

15/2022, 2022) acknowledges citizen communities as new types of actors, including in potential activities such as the production of energy efficiency services and charging for EVs without restriction on the energy vector utilized. Officially, the first community in Portugal was created in 2021 in Miranda do Douro; however, numerous initiatives had already been launched before, supported by European funding, such as the COMPILE, PROSEU, and COMSOLVE projects. In 2023, Portugal counted 4 operational REC and more than 350 requests for licensing, mainly in the central and northern parts of the country, in residential or industrial structures with the active involvement of local cooperatives, as C Coop Cooperative para a Sustentabilidade da Ilha da Culatra, which governed the case of Culatra as the *first fishing community with energy shared* among the members acting as an exemplary pilot project in the area in the combat of energy poverty of the region.

Conversely, the case of Italy presents significant particularities on the viewpoint of EC development. Through the transposition of the EU Directive (law Article 42/bis of the Decreto Milleproroghe (DL162/19) (Decreto-Legge Del 30/12/2019 No 162, 2019)), the Italian government defines the possibility of establishing collective self-consumption models and energy communities made up of different entities and equip themselves with infrastructure for the renewable production and the adoption of consumption patterns based on sharing activities and the introduction of social mechanisms to reinforce responsibility and activate participation of the community members in the stages of production, consumption, and energy exchanges. In 2022, Italy counted more than 70 REC projects and 16 collective self-consumption projects, mainly relying on photovoltaic, biogas, or thermal solar vectors, and, in lower proportions, wind power, while many other cases are in planning processes. Notable examples of Italian cases in the Ferla municipality (south-eastern Sicily) as part of the Multiply project (MULTIPLY, 2025) were led by several incentives, such as communication campaigns and incorporation into existing associations aiming at territorial cohesion and local economic savings.

Other cases, such as Greece, are more recent and require significant amendments to implement EC concepts and activities within their territories. Complexities in the legislative transposition, administrative and funding barriers, or social resistance are among the main difficulties. In the case of Greece, only in 2023 did a new era for the institutional framework ushered the importance of ECs applicability focusing on the multiple benefits to their members in the respective areas established in 2023 via the Law No. 5037/2023 (Law No. 5037 of 2023). Renaming the Energy Regulatory Authority to the Waste, Energy and Water Regulatory Authority and Expanding Its Scope with Responsibilities over Water Services and Urban Waste Management, Strengthening Water Policy—Modernisation of the Legislation on the Use and Production of Electricity from Renewable Sources through the Incorporation of EU Directives 2018/2001 and 2019/944, 2023). Among the ECs developed, 1669 are active (under Law No. 4513/2018 (Douvitsa, 2018)) in self-production or commercial configurations; the differences in numbers compared to other European countries are evident, considering that *Greece became the first country in the EU enabling virtual net-metering*; typical cases with diverse features and socio-technological innovation are (Delicado et al., 2017):

- ESEK (Karditsa Energy Community) (ESEK SYN PE, 2018): Established in 2010 as a profit citizen energy cooperative to foster renewable energy in the area under the Law No. 4513/2018 exploiting the potential of the agricultural character of the region with the great biomass potential, forestry, and wood processing to uptake as well the bioenergy technological contributions.
- Minoan Energy EC Crete awarded the European Sustainable Energy Award in 2022 in the Local Energy Action category (European Commission: Energy, Climate change, Environment, 2022), including over 100 households acting as a cooperative of individuals and local businesses with a pivoting role in the energy transition and energy democracy developed in an evolving network of partners and research collaborations for virtual net metering approaches.
- WenCoop EC, Thessaloniki (Energy Community Platform, 2021) as the first EC for women entrepreneurs in Greece and Europe, with 66 members active since 2021 in an all-women cooperative. The EC operates a 1 MW solar power plant in the areas of Kassandreia and Halkidiki with the ambition of being expanded in three more photovoltaic parks for electromobility, and other related to production activities, mainly to provide services to vulnerable families, and especially to single-parent families of women.

From the above analysis, benchmarking approaches clearly show differences in the transposition and practicalities of ECs across the diverse MS in Europe, regarding the organizational forms, or the understanding and application of the concepts, such as the collective self-consumption, which in most cases does not include specific requirements of legal entities, but it is more about agreements between the participating actors; with particular exceptions for France, or Brussels-Capital Region. Other observations are linked with the governance structures or the proximity criterion for the particular case of RECs' applicability and the effective control over the renewable potential. In some MS, spatial restrictions are applied as rules for REC development, whereas CECs do not have per se, geographical or system-related scope limitations (e.g., RED II (European Commission, 2023) refers to the "same building or multi-apartment block," nonetheless, in France, for instance, self-consumption is limited in its spatial expansion) (Frieden et al., 2019).

References

A. Agrawal, K. Hockerts, Impact investing: review and research agenda. J. Small Bus. Entrep **33**, 153 (2021)

J. Barlow, Wiltshire wildlife community energy. Annu. Rep., 2022–2023 (2024) https://wwce.org/wp-content/uploads/2024/04/WWCE-Annual-report-2023_V2.pdf

T. Bauwens, Polycentric governance approaches for a low-carbon transition: the roles of community-based energy initiatives in enhancing the resilience of future energy systems, in *Complex Systems and Social Practices in Energy Transitions*, (2017)

T. Bauwens, B. Gotchev, L. Holstenkamp, What drives the development of community energy in Europe? The case of wind power cooperatives. Energy Res. Soc. Sci. **13**, 136–147 (2016)

S. Becker, C. Kunze, Transcending community energy: Collective and politically motivated projects in renewable energy across Europe. People Place Policy, 180–191 (2014)

A.L. Berka, Community renewable energy in the UK: a short history, in *Handbuch Energiewende und Partizipation*, ed. by L. Holstenkamp, J. Radtke, (Springer, Wiesbaden, 2018)

A.L. Berka, E. Creamer, Taking stock of the local impacts of community owned renewable energy: A review and research agenda. Renew. Sust. Energ. Rev. **82**, 3400–3419 (2018)

S. Bouzarovski, N. Simcock, Spatializing energy justice. Energy Policy **107**, 640–648 (2017)

T. Braunholtz-Speight, M. Sharmina, E. Manderson, C. McLachlan, M. Hannon, J. Hardy, S. Mander, Business models and financial characteristics of community energy in the UK. Nat. Energy **5**, 169–177 (2020)

A. Caramizaru, A. Uihlein, *Energy Communities: An Overview of Energy and Social Innovation* (JRC Science for Policy Report, European Commission, 2020) https://publications.jrc.ec.europa.eu/repository/handle/JRC119433

E. Christiansen. *ELENA completed project factsheet. Avedore Green City (Denmark).* (2023). https://www.eib.org/files/elena/87-project-factsheet-agc.pdf

Climate Change Laws of the world. *The environmental activities decree of The Netherlands* [Letter]. (2018). https://climate-laws.org/document/the-environmental-activities-decree-of-the-netherlands-besluit-activiteiten-leefomgeving-bal_3577

Coopératives d'énergie renouvelable, citoyenne et solidaire. (2020, June 12). Enercoop. https://www.enercoop.fr/

S.L. Cutter, B.J. Boruff, W.L. Shriley, Social vulnerability to environmental hazards. Soc. Sci. Q. **84**, 242–261 (2003)

Decree Law 15/2022. https://diariodarepublica.pt/dr/en/detail/decree-law/15-2022-177634016. (2022)

Decreee Law No 162/2019. *Republic Diary no 206/2019* (2019). https://diariodarepublica.pt/dr/en/detail/decree-law/162-2019-125692189

A. Delicado, R. Iglesias, M. Ghislanzoni, M. Truninger, C. Pereira, R. Roque, J. Sa Couto, G. Macca, C. de Benedictis, V. Ferreira, A. Horta, T. Istoriou, C. Felix, M.J. Prados, *Planning and engagement arenas for renewable energy landscapes-PEARLS_Marie Sklodowska Curie Actions (MSCA) Research and Innovation Staff Exchange (RISE). WP5 Report: Renewable Energy Communities in Portugal* (Spain, Italy and Greece, 2017)

Deltawind. (2021). *Deltawind cooperative*. https://deltawind.nl/

I. Douvitsa, The new law on energy communities in Greece. Cooperativismo e Economía Social (CES) (2018)

A. Dudka, N. Magnani, G. Koukoufikis, Analysing perspectives on capital, mutual, and general interest: A comparative study of energy cooperatives in Belgium and in Italy. Energy Res. Social Sci. **116** (2024)

Ecopower. *A citizens' cooperative for renewable energy (REScoop)*. (2025). https://www.ecopower.be/over-ecopower/onze-werking

enercoop. *Coopératives d'énergie renouvelable citoyenne et solidaire*. (2025). https://www.enercoop.fr/

Energ4All. *E4A Co-ops. Green Cooperative Energy*. (2025). https://energy4all.co.uk/e4a-co-ops/

Energie Partagée. *Rapport annuel 2020. L'énergiee par les citoyens pour les citoyens!*. (2020). https://energie-partagee.org/wp-content/uploads/2021/05/Energie-Partagee-Rapport-dactivite-2020.pdf

Energy Community Platform. *WEnCoop energy cooperative*. (2021). https://energycommunityplatform.eu/communities/energy-community-wencoop/

ESEK SYN PE. *Energy Community of Karditsa (ESEK)*. (2018). https://www.esek.gr/en/archiki-english/

European Commission. (2023). Renewable energy directive (RED II) directive (EU) 2018/2001 (recast) on the promotion of the use of energy from renewable sources.. https://www.europex.org/eulegislation/renewable-energy-energy-directive/

European Commission: Energy, Climate change, Environment. *Energy community in Crete wins European Sustainable Energy Award for paving the way for free electricity for all.* (2022). https://sustainable-energy-week.ec.europa.eu/energy-community-crete-wins-european-sustainable-energy-award-paving-way-free-electricity-all_en

EWS. (2016). *Elektrizitätswerke Schönau—Who We Are. Elektrizitätswerke Schönau.*. https://www.ews-schoenau.de/

Flexalen. *Eco-village Permatopia Rolling out a Sustainable Future with Flexalen. Karis Permatopia.* (2018).. https://thermaflex.com/case-studies/a-sustainable-future-with-flexalen/

D. Frieden, A. Tuerk, J. Roberts, S. D'Herbemont, A. Gubina. *Collective Self-Consumption and Energy Communities: Overview of Emerging Regulatory Approaches in Europe. Compile: Integrating Community Power in Energy Islands.* (2019). https://www.rescoop.eu/uploads/rescoop/downloads/COMPILE_Collective_self-consumption_EU.pdf

M. Gancheva, S. O'Brien, N. Crook, C. Monteiro, *Models of Local Energy Ownership and the Role of Local Energy Communities in Energy Transition in Europe* (European Committee of the Regions, 2018)

A. Garvey, J.B. Norman, M. Büchs, J. Barrett, A 'spatially just' transition? A critical review of regional equity in decarbonisation pathways. Energy Res. Social Sci. **88** (2022)

L. Gorroño-Albizu, K. Sperling, S. Djorup, The past, present and uncertain future of community energy in Denmark: critically reviewing and conceptualizing citizen ownership. Energy Res. Social Sci. (2019)

E.M. Gui, M. Diesendorf, I. MacGill, Distributed energy infrastructure paradigm: community microgrids in a new institutional economic context. Renew. Sust. Energ. Rev., 1355–1365 (2017)

D. Hendricks, R. Mesquita, PV prosumer guidelines for eight EU member states. Eur. Renew. Energies. (2019) https://www.pvp4grid.eu/wp-content/uploads/2019/05/1904_PVP4Grid_Bericht_EUnat_web.pdf

R. Hewitt, N. Bradley, A.B. Compagnucci, C. Barlagne, A. Ceglarz, R. Cremades, M. McKeen, I.M. Otto, B. Slee, Social innovation in Community Energy in Europe: a review of the evidence. Front. Energy Res. (2019)

L. Holstenkamp, F. Kahla, What are community energy companies trying to accomplish? An empirical investigation of investment motives in the German case. Energy Policy **97**, 112–122 (2016)

B. Huybrechts, S. Mertens, The relevance of the cooperative model in the field of renewable energy. Annal. Public Cooperat. Econ. **85**, 193–212 (2014)

M. Koltunov, S. Pezzutto, A. Bisello, G. Lettner, A. Hiesl, W. Van Sark, A. Louwen, E. Wilczynski, Mapping of Energy Communities in Europe: Status quo and review of existing classifications. Sustainability **15** (2023)

H.J. Kooij, M. Oteman, S. Veenman, K. Sperling, D. Magnusson, J. Palm, F. Hvelplund, Between grassroots and treetops: Community power and institutional dependence in the renewable energy sector in Denmark, Sweden and The Netherlands. Energy Res. Soc. Sci. **37**, 52–64 (2018)

J. R. Laurel. *Regulatory frameworks and community engagement strategies: A comparison of Belgian policy against Denmark and The Netherlands* [Master thesis, UHasselt. Faculty of Business Economics]. (2024). https://documentserver.uhasselt.be/bitstream/1942/43971/1/6b65d81b-e0b1-4a26-a449-62fe7a69690d.pdf

Legambiente. *Comunità Rinnovabili. Legambiente*. (2022). https://www.legambiente.it/wp-content/uploads/2021/11/Comunita-Rinnovabili-2022_Report.pdf

F. Mey, M. Diesendorf, Who owns an energy transition? Strategic action fields and community wind energy in Denmark. Energy Res. Soc. Sci. **35**, 108–117 (2018)

M. Montojo, Crevillent—Spain's first local energy community. One Earth. (2020) https://www.oneearth.org/crevillent-spains-first-local-energy-community/

S. Moroni, V. Alberti, V. Antoniucci, A. Bisello, Energy communities in the transition to a low-carbon future: a taxonomical approach and some policy dilemmas. J. Environ. Manag. **236**, 45 (2019)

MULTIPLY. *Multiplying feasible microalgae-derived products at scale* [Research project]. (2025). https://cordis.europa.eu/project/id/101214894

J. Nikolic, O. Clemmensen, M. Gerritsen, K. Sperling, H. J. Kooij. *COPPER project: energy communities: Inventory of barriers and opportunities and best practice*. (2025). https://dutpartnership.eu/sites/default/files/2025-11/COPPER%20ENERGY%20COMMUNITIES%20INVENTORY%20OF%20BARRIERS%20OPPORTUNITIES%20AND%20BEST%20PRACTICE.pdf

Nordic Energy Research. (2023). Energy communities. https://pub.norden.org/nordicenergyresearch2023-03/netherlands.html

Padovan, D., Arrobio, O., Sciullo, A., Gilcrease, W., Gregg, J. S., Henfrey, T., Wierling, A., Schwanitz, V. J., Labanca, N., Dunlop, T., Alvarez Polo, L., & Candelise, C. (2019). *Collective Action Initiatives. Some Theoretical Perspectives and a Working Definition*

A. Parreño-Rodriguez, A.P. Ramallo-González, M. Chinchilla-Sánchez, A. Molina-García, Community energy solutions for addressing energy poverty: A local case study in Spain. Energy Build **296** (2023)

J.P. Petersen, Energy concepts for self-supplying communities based on local and renewable energy sources: a case study from northern Germany. Sustain. Cities Soc. **26**, 1–8 (2016)

Primer Modelo de Autoconsumo Comunitario - Red Eléctrica de España (2021). https://www.ree.es/es/sostenibilidad/proyectos-destacados/innovacion-social/primer-modelo-autoconsumo-comunitario

I.F.G. Reis, I. Gonçalves, M.A.R. Lopes, C. Henggeler Antunes, Business models for energy communities: A review of key issues and trends. Renew. Sust. Energ. Rev. **144** (2021)

REScoop. *European Federation of Renewable Energy Cooperatives*. (2016). https://www.rescoop.eu/

C. Romero-Rubio, J.R. de Andrés Díaz, Sustainable energy communities: A study contrasting Spain and Germany. Energy Policy **85**, 397–409 (2015)

N. Rossetto, S.F. Verde, T. Bauwens, A taxonomy of energy communities in liberalized energy systems, in *Energy Communitie*, (Academic Press, Cambridge, 2022), pp. 3–23

C. Sebi, A. Vernay, Community renewable energy in France: The state of development and the way forward. Energy Policy **147** (2020)

N. Simcock, R. Willis, P. Capener, *Cultures of Community Energy. International Case-studies* (British Academy, 2016)

Sweco. *Distributed Electricity Production and Self-Consumption in the Nordics. A Report to the Nordic Council of Ministers*. (2019) https://www.nordicenergy.org/wordpress/wp-content/uploads/2019/06/Distributed-energy-production-and-self-consumption-20190607-1.pdf

D. Van der Horst, Social enterprise and renewable energy: emerging initiatives and communities of practice. Soc. Enterp. J. **4**, 171–185 (2008)

A.L. Vernay, C. Sebi, Energy communities and their ecosystems: a comparison of France and The Netherlands. Technol. Forecast. Soc. Chang. **158**, 120123 (2020)

A.L. Vernay, C. Sebi, F. Arroyo, Energy community business model and their impact on the energy transition: lessons learnt from France. Energy Policy **175** (2023)

M. Weber, *Methodology of Social Sciences*, 1st edn. (Routledge, 2011)

A. Wierling, V.J. Schwanitz, J.P. Zeiß, C. Bout, C. Candelise, W. Gilcrease, J.S. Gregg, Statistical evidence on the role of energy cooperatives for the energy transition in European countries. Sustainability **10** (2018)

A. Wierling, V.J. Schwanitz, J.P. Zeiss, C. von Beck, H.A. Paudler, I.K. Koren, T. Kraudzun, T. Marcroft, L. Müller, Z. Andreadakis, C. Candelise, S. Dufner, M. Getabecha, G. Glaase, W. Hubert, V. Lupi, S. Majidi, S. Mohammadi, N. Safara Nosar, et al., A Europe-wide inventory of citizen-led energy action with data from 29 countries and over 10000 initiatives. Sci. Data **10** (2023). https://www.nature.com/articles/s41597-022-01902-5

Zeeuwind coop. *Zeeuwind*. (2021). https://www.zeeuwind.nl/

Chapter 5
Navigating Belgian Energy Communities: Challenges and Opportunities

5.1 Energy Challenges and Opportunities in Belgian Territories: What and Why?

In their study of energy consumption trends across various sectors, Singh et al. (Kumar Singh et al. 2013) identified the distinct case of Belgium compared to that of EU-27 with the industrial sector consuming most of the primary energy. Energy consumption for the building sector accounts for more than 70%, higher than the European average (Energy Annual Statistics 2010). Considering the territorial particularity in Belgium into three regions, namely Flemish, Brussels capital, and Walloon region (Renard et al. 2008), the transposition of the European Directives in energy efficiency differentiates in a number of incentives related to regulatory adaptations, training, and information and financial mechanisms (Kumar Singh et al. 2013).

A lack of initiative from local stakeholders and supporting mechanisms, along with associated institutional and legal constraints, creates structural obstacles to energy community applications. The most important challenge in Belgium remains the regional government structure, with a heterogeneous legal framework that sets out different directives for the three federations; each of them has adopted its unique approach to transposing RSC, REC, and CEC definitions at designated timelines.

In particular, in the case of Wallonia, the adoption of the REDII (European Commission 2023b) defines the RECs purpose in the production, consumption, and storage for the public benefits with further balances of consumption and production flows within a local perimeter of energy-sharing, in which every natural person, local authority, or company participates in the process. On its side, the Flemish government draws on the concept in the areas of production, distribution, consumption, supply, storage, and electromobility by introducing the idea of "active consumer," whereas in the Brussels-Capital Region, the concept of "self-consumption" is primordial.

S. Koutra, *Energy in the Community*, SpringerBriefs in Applied Sciences and Technology, https://doi.org/10.1007/978-3-032-27148-8_5

Flanders drafted and adopted a legislative framework in December 2020 for all three types of ECs mentioned. As for Wallonia, collective self-consumption had decreased in 2018 and 2019; however, there was framework legislation in the decree for REC definition transposition and no legislation for CECs. Brussels has adopted the 2018 definition for collective self-consumption and transposed the EU definitions for REC and CEC by 2021 (RESCOOP.EU 2025b).

Looking specifically at energy community barriers, the analysis for the Belgian frame identifies *a lack of experience and expertise in the community models at social, cultural, and organizational levels*. The challenges are multiple, starting with transparency and trust among prosumers, the complex legal and funding support schemes, and the lack of prior knowledge.

5.2 National Objectives Toward Energy Transition, Constants in Belgian Energy Plans, and Trends and Transposition from the European Legal Framework

Belgium, like other EU MS, has integrated EC concepts into its national regulatory framework to accelerate decarbonization and energy transition strategies and achieve sustainable development across its territories. Belgium supported the initiative by providing financial, regulatory, and technical assistance and by framing the energy and climate policies for its regional federations. Each federation outlined the territorial, political, and cultural specifics for energy-efficiency improvements and the promotion of local energy generation, with a corresponding reduction in requirements. Although the priorities seem similar, significant differences are observed across the three regions.

5.2.1 Regulatory Framework for Energy Communities in Flanders

There is no clear definition of a renewable self-consumer community in Flanders, nor specific rules for communities on EC activities, such as generation, consumption, storage, or selling, with the only exception being that the exchange of surplus electricity directly between prosumers is not allowed. Conversely, CECs become possible due to the "Energy Performance Certificates"[1] as supporting mechanisms for self-consumers to increase capacity investments in solar panels (Inês et al. 2020).

[1] Policy instrument that not only certifies a building's energy efficiency but also seeks to incentivize investment in greener properties, in a multiple indicator study. https://build-up.ec.europa.eu/en/resources-and-tools/articles/energy-performance-certificates-boosting-market-energy-efficient, Accessed on: 16/02/2026

Flanders has developed the Flemish Energy Decree (Flemish Government 2010), which includes specific provisions for EC support and emphasizes open and voluntary participation, autonomy, and effective control strategies (Laurel 2024). The Energy Decree of 19th November 2010 (Flemish Government 2010) established a comprehensive legal framework for energy policies and management, including standards for production, distribution, supply, and consumption, and outlined operational guidelines for the governance structure of the ECs and their organization. The same decree also emphasized the ECs' development and integration, defining them as citizen or renewable, and sets out guidelines for their practical operation, including the sharing rules, and detailed narratives about their monitoring, reporting, and accountability (Laurel 2024); in this lens, a specific subsidy program introduced by the Flemish Energy and Climate Agency (VEKA), the "Call Groene Stroom" (Flanders 2021), offers financial support for new PV and wind turbines installations, including provisions for ECs development to support local renewable generation boosting the ECs scalability in the region and facilitate their efficient operations.

The Flemish government states that ECs carry out activities relevant to the areas of production, distribution, consumption, supply, aggregation, sharing, and other relevant and proposes the introduction of local flexibility markets and the concept of "active consumer"[2] who can store, participate, sell, or act as a provider in flexibility provision or aggregation activities and determines specific geographic criteria related to the resources proximity. In the same regulation, Flemish Government explicitly develops the relevant business plan of the EC development for the investigation of distribution networks within the specified areas (Frieden et al. 2019).

5.2.2 Regulatory Framework for Energy Communities in Wallonia

On the other side, Wallonia developed a context-specific framework for Collective Self-Consumption (CSC), adopted in October 2018, with a definition of RECs in reference to the REDII (European Commission 2023b), which clearly identified the specific scopes of production, consumption, storage, sale of renewable electricity, and all the common benefits for participants. The same law also defined the "local perimeter" as "technically, socially, and environmentally" optimal for promoting self-consumption (Frieden et al. 2019).

Enacted in April 2001, the Walloon Energy Decree (Walloon Parliament 2014), amended and in compliance with EU Directives, established a relevant regional framework for ECs' operational guidelines and actors' responsibilities for community engagement, to adhere to principles for a democratic and transparent

[2]According to the Electricity Market Design (EMD) (European Commission 2024), an "active consumer" is a final user (customer) (or group of them) who consumes or stores electricity generated within confirmed boundaries, sells self-generated electricity, or participates in flexibility services

framework and to support measures (e.g., subsidies for renewable energy systems installations, e.g., solar panels, wind turbines, and biomass) that foster a decentralized and community-driven transition (Laurel 2024).

5.2.3 *Regulatory Framework for Energy Communities in Brussels*

While not a specific outline for EC operations, the Brussels Energy Transition Plan (2018–2030) (Région de Bruxelles Capitale 2019) promotes energy practices for decentralized energy production, renewable energy use, and enhanced citizen participation to create a favorable ecosystem for EC development and growth.

Primarily established by the Energy Communities Ordinance, ECs are outlined as well by supporting mechanisms for their further development, and in particular (Laurel 2024).

- ECs are legal entities based on open and voluntary participation, with the primary objectives being the environmental and social benefits for their participants.
- ECs' activities are summarized in generation, consumption, storage, sharing, and sale to encourage flexible services.
- The Brussels government financially supports the EC's replicability with relevant contributions, e.g., grants, loans, subsidies, or technical assistance (e.g., guidance and relevant expertise), and engages community with specific decision-making processes.

5.3 Operational Levels of Energy Communities in Belgian Regions: Process of Establishment and Energy-Sharing Models

Belgium's energy and climate policy matters are undertaken by the federal government and the regional governments. The former is responsible for large-scale electricity generation and transmission, and the latter oversees distribution. Hence, the relevant legislation is established in every Belgian region independently (European Commission 2023a). Any energy community is obliged by the respective Energy Regulator to be granted permission to initiate its activities and to comply with the standards and conditions set by the same overseeing body, for Brussels it is Brugel, for Flanders it is VREG, and for Wallonia it is CWaPE (Commission Wallonne Pour Energie) (Frieden et al. 2020).

The implementation of EU directives on Energy Communities is managed at the regional level. In the Flemish Region, the Flemish Energy Agency (VREG) provides guidelines and support for establishing ECs, with a focus on renewable energy projects and local energy initiatives. In the Walloon Region, the regional government has introduced measures to promote ECs, including financial incentives and regulatory frameworks to facilitate the development of renewable energy projects at the community level, and CWaPE is responsible for registering ECs. Similarly, in the Brussels-Capital Region, Sibelga offers support for local energy initiatives, emphasizing sustainability and community involvement in energy projects.

5.3.1 Process of Establishment

Each Belgian region has established its own procedures for setting up ECs. While these procedures may be distinct in detail, they share similar underlying principles (CWAPE 2019); this chapter focuses on the Brussels-Capital Region as a representative case, illustrating the process based on the framework provided by the official body "Bruxelles Environnement." This regional procedure was selected for demonstration due to its clarity and accessibility. The establishment of an EC in Brussels-Capital Region follows two main phases:

5.3.1.1 Planning, Design, and Implementation Phase

This phase (Fig. 5.1) includes the EC project's definition related to the scope, objectives, financing structures, the consumption profiles of the potential members, and a suitable perimeter. Out of the four (4) options set by Brugel and how electricity will be distributed between members, then determination of the price for the electricity to be shared, followed by administrative procedures (e.g., establishing statutes governing the internal activities for the EC selecting an appropriate legal form, setting up a convention of all signatory members to secure the right and conditions of exiting the EC and the procedures to take place in case of non-payment by one of the member), and admission of the required documents to the Brugel). Once approved, the authorization to commence the activities is granted and valid for ten (10) years, renewable at the end of its term. Additionally, the status of the EC will be published in the public database of Brugel. The final step in this phase is implementing the RES assets.

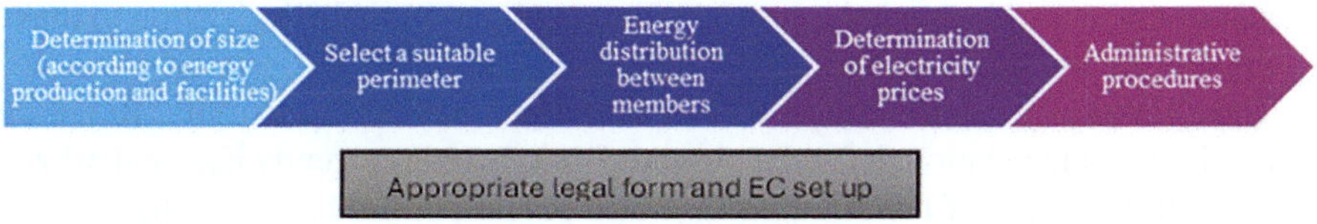

Fig. 5.1 Process of EC planning, design, and implementation. (Koutra, 2026)

5.3.1.2 Management Phase

Once operational, the community enters the management phase, which involves overseeing energy distribution, managing billing for community members and network usage under Sibelga's responsibility, and handling procedures for members joining or leaving the REC, in accordance with the statutes established in the first phase.

5.4 Operational Levels of ECs in Belgian Regions

5.4.1 Renewable Self-Consumption (RSC)

The REDII (European Commission 2023b) defines the "self-consumer" as "the customer [...] who generates electricity for his own consumption, who may store or sell self-generate" and consists of an emerging concept in most MS.

In addition to sharing surplus electricity with nearby citizens, individuals are encouraged to maximize their own energy self-consumption, referred to as "renewable self-consumers," to consume electricity generated by themselves from renewable energy sources, as rooftop solar PV systems or nearby installations.

In the context of EC, the definition for these prosumers was formally introduced in RED II, particularly in Article 2(15) of RED II, it defines: "Renewable self-consumer means a final customer... who generates renewable electricity for their own consumption, and who may store or sell self-generated renewable electricity."

5.4.2 Flanders

In Flanders, the REC and CEC concepts are converted at the local level, through the Energy Decree (set of regulations and policies imposed by the Flemish government to control the energy sector regionally) established by VREG. Both REC and CEC must declare the rules encompassing the control of its members or associates. This legislation also puts forward the oversight and monitoring of the activities of both REC and CEC to keep VREG up to date regarding the activities they conduct and if there is any change induced:

- Their composition and how they plan to meet their objectives (voluntary entry, autonomy, control, ownership, etc.), with no obligation in the legislation to make this information transparent.
- The community's definition of technical and geographical proximity. Based on this Flemish legislation, it can be observed that there is currently no standardized or clear definition of the criterion of geographical proximity within the Flemish

ECs; instead, its interpretation largely depends on the community's understanding of the concept.

However, the dissimilarities in REC and CEC regulations encompass a wide range of factors:

For CEC:

The members of CEC should reach an agreement with the community and state their rights and obligations, if they share energy. Additionally, the control of the activities within the community is effective only by either natural persons, local authorities, or small enterprises, large commercial companies whose energy sector is their field of activity, and targeting profits as their primary objective of participation in the community are excluded.

For REC:

- RECs' activities are concentrated on energy production, self-consumption, energy selling, and sharing.
- Bound to limit the participation of members based on technical and geographical proximity.
- Entitled to the ownership rights of the RE assets and has the right to carry out its energy activities.
- Emphasizes energy autonomy.

For an EC to operate in Flemish territory, it is subject to an electricity supply license (Article 4.3.1 of the Energiedecreet), granted by the VREG; this requirement is applicable for electricity supplied through a distribution network or a local transport network to customers.

While Flanders is the Belgian frontrunner in fostering the most advanced energy cooperatives (such as Ecopower) (Felice et al. 2021), there is still lack of local legal frameworks and tariffing structures that facilitate the economic feasibility of RECs. A study done by Felice et al. suggests that the inclusion of various flexible energy sources is the primary solution for making RECs feasible economically in Flanders, the best case highlighted a 17% cost reduction compared to the business-as-usual (BAU) case due to the capacity tariff which these technologies possess that can take advantage of the peak shaves (Felice et al. 2021).

In November 2020, the Flemish Government published a draft legal framework to transpose the provisions of IEMD and, to a limited extent, those of REDII. This framework outlines the scope of activities that ECs may undertake, including electricity production, distribution, consumption, supply, aggregation, flexibility services, energy sharing, storage, electric vehicle charging, energy efficiency services, and other energy-related services. Notably, the draft also introduces the concept of local flexibility markets. The legislation merges the EU-recognized roles of "active consumer" and "renewable self-consumer" into a single category—active consumer. However, it does not explicitly define a separate category for "jointly acting renewable self-consumers." Within the proposed framework, residents of a shared building, such as an apartment complex, are allowed to collectively use locally generated energy.

An active consumer, however, may:

- Store energy
- Participate in the provision or aggregation of energy services
- Operate as a flexible service provider
- Sell energy, including through Power Purchase Agreements (PPAs)

While the transposition of RECs and CECs generally aligns with EU legislation, the Flemish framework introduces context-specific provisions; for example, participation in an REC is linked to technical or geographic proximity, and the Flemish Government is authorized to define these proximity criteria based on the energy community's objectives.

Moreover, the Flemish Government has mandated a cost-benefit analysis to be conducted by the energy regulator VREG to assess the impact of energy communities on the distribution network, including potentially avoided infrastructure investments. Furthermore, under certain conditions, ECs may also be authorized to operate their own distribution networks within designated areas.

The Flemish Energy Decree also conceives ECs as a "single concept" but acknowledges minor distinctions between REC and CEC definitions in its transposition. VREG is responsible for registering, monitoring, and ensuring compliance with ECs. Nonetheless, it also has other energy-sharing models that are excluded from the transposition, namely renewable self-consumption (RSC), peer-to-peer (P2P) energy trading, and collective self-consumption (CSC).

Similar to the classification used by Brugel in the Brussels-Capital Region, VREG recognizes those but adds fourth model of RSC. In contrast to Brugel, which differentiates ECs into LECs, RECs, and CECs, VREG does not apply such distinctions within the operational framework of ECs' in Flanders.

According to VREG's database of energy-sharing projects and its December 2024 reporting, the RSC energy-sharing model accounts for the largest number of implemented projects in Flanders, with 2773 cases, followed by P2P energy trading (1212 projects), CSC (148 projects), and ECs (148 projects) and 24 operational RESCoops.

The Energy Sharing Typologies (VREG)

- **Renewable Self-Consumption (RSC)**: Under the heading of "energy sharing with yourself" or "titular sharing," a person, company, or (local) government can use self-produced green energy in all its residences, branches, or buildings.
- **Pair-to-Pair Energy Trading with a Person or an Organization**: Citizens, companies, and governments share their self-produced green energy among natural persons, as well with an organization, a company, or a government.
- **Collective Self-Consumption (CSC) in an Apartment Building or Other Communal Building**: "Energy sharing in a communal building" makes it possible to share the self-produced green energy with the residents of an apartment building or the users of a "multifunctional" building.
- **Energy Communities (ECs)**: These include REC and CEC transpositions. Also known in Flemish as "Energiegemeenschap van burgers" (EGB) and

Hernieuwbare-energiegemeenschap HEG, respectively. Citizens, businesses, or (local) authorities can form an energy community and invest together in green energy and efficiency.

As far as the legislative framework is concerned, regional pieces of the Flemish Codex (2025) include energy-sharing models (Article 7.2.1), P2P (Article 7.2.2), and active customers (Article 4.4.2) with continuous efforts for transposition into local legislations (RESCOOP.EU 2025a) and assist local actors and potential participants through subsidies and other forms of contribution to establish ECs, especially those tackling energy poverty.

At the municipal level, the Energiehuis (Energy House) was introduced, their primary goal is to empower citizens, communities, and other stakeholders to actively participate in energy efficiency efforts achieved through various forms of support, including loans, informational guidance, and technical advice. Energy Houses assist individuals and groups in undertaking actions such as reducing energy consumption, installing photovoltaic systems, or carrying out building renovations aimed at improving energy performance. At the federal level, a more engaging solution is the promotion of offshore wind farms and enhancing ECs capacities to engage and finance their own wind farm projects (Frieden et al. 2020).

5.5 Brussels

Brussels-Capital Region has transposed the EU definitions of ECs in the regional energy regulations (e.g., REC and CEC); however, all active communities so far are still registered as Local Energy Communities. This approach seemed more suitable, as these communities were established before the release of the RED II and IEMD directives. Brussels-Capital Region has an overall positive assessment of the translation of REC and CEC definitions into national regulations, according to the Transposition tracker designed by RESCoop.eu to monitor and evaluate the progress of EC definitions in European Member States. To transpose the CEP provisions synergistically, Brussels Ordinance has established four (4) definitions for ECs, namely (energy community, citizens energy community, renewable energy community, and local energy community), for which Brugel evaluates and validates the legal status and regulatory compliance, and Sibelga manages the technical and consumption aspects of the shared energy flows. All energy communities share the attributes of being objective, transparent, non-discriminatory, and autonomous; however, they differ in their effective control, eligibility criteria for participation, and ownership rights. The Energy Regulator Brugel does not specify a legal form for creating an EC; thus, EC members are free to adopt a suitable legal form, but they are subject to certain conditions (RESCOOP.EU 2022).

Brugel grants valid permission for the EC to carry out its operations for ten (10) years, renewable, publishes the list of the permitted ECs, and describes their activities but also terminates the authorization of EC establishment if it no longer meets the set criteria or does not follow its obligations. Therefore, Brugel grants, renews, or withdraws authorization for energy communities or examines the removal of any unjustified obstacles hindering the development of ECs. DSO (in this case, Sibelga) is tasked with providing online support for energy-sharing members to access their consumption data and install smart meters free of charge. Brussels' legislation emphasizes renewable energy sources in energy sharing via ECs and promotes favorable tariffs for the sharing of electricity from RES that takes into account the existing distribution network (Bruxelles environnement 2023).

Nevertheless, there are no supportive policies or measures to promote participation in ECs' incentives by low-income and vulnerable households. However, Energie Commune, as a facilitator, organizes participatory activities (e.g., workshops) focused on how to undertake energy sharing in social housing (Williams 2023). Additionally, Energie Commune undertakes other responsibilities as well, in the framework of promoting ECs, such as providing free advice to EC project promoters, and building capacity for local authorities who need guidance on the aspects of regulations.

The regional energy regulator has permitted certain exceptions aligned with the provisions of CEP. These exceptions have enabled the establishment of several pilot projects related to energy communities, with Sibelga actively participating in most of these initiatives. Additionally, the regional government has provided subsidies to the renewable energy association of the Wallonia-Brussels Federation to support citizen-led pilot projects.

Besides transposing EC definitions, Brussels-Capital Region has developed a structured framework for energy sharing based on two more models, each varying in scale, complexity, and regulatory requirements, designed to promote citizen participation in energy transition, and supported by the regional energy regulator Brugel and DSO Sibelga. There is no requirement for these models to possess a supply license to share energy with members; however, there must be third party responsible (RESCOOP.EU 2022):

1. **Peer-to-Peer (P2P) Energy Trading (Échange de pair à pair)**: P2P trading is the most straightforward form of energy sharing in Brussels, allowing two active individual customers to directly exchange surplus electricity. Key features of this model include:
   - **Bilateral Agreement**: The electricity price is freely negotiated between the participants with agreements on energy sharing and costs.
   - **Optimized Consumption Matching**: To maximize benefits, Sibelga recommends pairing participants with complementary consumption profiles, improving overall energy self-consumption.
   - **Facilitation**: "Bruxelles Environnement" may assist in initiating P2P projects as a facilitator.
   - **Eligibility Conditions**:
     - The electricity must be generated from renewable sources.
     - Both participants must be active customers, equipped with a smart meter.

- A contractual agreement must be signed outlining the terms of exchange.
- One party (the producer or their representative) must act as the single point of contact (SPOC) with Sibelga.
- No legal entity or organization is required to participate.
- Third-party investors are permitted, as long as energy ownership remains with the client.

  Data management and billing can be delegated to a third party without them becoming an active customer.

2. **Collective Self-Consumption (CSC) (Clients actifs agissant conjointement):** This model enables residents within the same apartment building to collectively use surplus renewable energy, typically from rooftop photovoltaic installations and is intended for groups of neighbors who want to locally share energy without injecting it into the wider grid.

5.6 Status Quo of Energy Communities Projects in Belgium

In 2022, only one peer-to-peer (P2P) project existed, involving just two participants. By March 2025, the total number of participants across all energy-sharing models had reached 1555, and by May 2025, this number had increased further to 1652 participants engaged in 207 energy-sharing projects, collectively generating 11.83 MW of renewable energy capacity. To monitor regional developments, the regulator Brugel provides real-time updates on EC data via its Microsoft Power BI dashboard. Among the typologies, CSC is the most dominant in most Brussels municipalities. Only three out of the 17 municipalities—Auderghem, Woluwe-Saint-Pierre, and Ganshoren—have a higher share of P2P projects. Schaerbeek hosts the most energy-sharing projects, with 26 initiatives, whereas Ganshoren has only four. Chronologically, a single P2P project was launched in 2022 followed by 49 new projects in 2023, nearly evenly split between P2P (24) and CSC (23). In 2024, CSC projects surged, reaching 71 initiatives, nearly triple the number recorded the previous year, solidifying CSC as the leading energy-sharing typology in the region (Brugel. Le regulateur bruxellois pour l'énergie 2025).

In Wallonia, the concept of "energy community" was initially brought to light through business parks, to highlight their potential. The Decree of the Walloon Region transposes the CEP definitions of REC and CEC into its legislation in 2018, followed by a transposition of the REC definitions in the Decree Law in 2019, and entered into force in 2022, acknowledging both RSC and CSC (CWAPE 2025b).

In accordance with RED (EU) 2018/2001(Directive (EU) 2018/2001 of the European Parliament and of the Council of 11 December 2018 on the Promotion of the Use of Energy from Renewable Sources (Recast) (Text with EEA Relevance.), 2018), the Walloon Electricity Decree enables collective energy sharing among active customers located within the same building. The decree has clearly delimited the definition of a building.

- Definition of "Building":[3]

  A building is defined as any fixed, covered, and enclosed immovable structure in which both the active energy customers and the installation producing electricity from renewable sources are established or situated.

This legal interpretation broadens the opportunities for renewable electricity production within collective buildings by allowing the use of additional surfaces, while maintaining compliance with the spatial limitations imposed by the EU directive, while the definition of REC refers to the geographical criterion as the "local perimeter," defined as: "A grid segment whose connection points are located downstream of one or more stations of public electricity transformation of medium and/or low voltage."

5.6.1 Relevant Walloon Legislation on Energy Communities

Decree of 5 May 2022 (CWAPE 2022): This decree constitutes a partial transposition of the EU Clean Energy Package into Walloon legislation (Frieden et al. 2020), particularly concerning the rights and roles of prosumers and energy communities (Service Public de Wallonie. Service Public Fédéral. Justice 2022).
Walloon Government Order of 17 March 2023 (Wallex 2023): This order provided the detailed regulatory framework for the operation of energy communities and energy sharing in Wallonia and provides information related to eligibility criteria, proximity conditions, authorization procedures, and the rights and obligations of participants (Service Public de Wallonie. Service Public Fédéral. Justice 2023).

The Walloon government retains the authority to review and amend the eligibility criteria for participants. The establishment of an REC requires formal authorization from both the DSO and CWaPE as the regulatory authority, with the main DSOs responsible for electricity and gas networks grouped under ORES (Opérateur des Réseaux Énergétiques de Service public), which is the primary DSO for most of Wallonia (covers approximately 75% of the territory). The law provides supportive measures, developed in coordination with DSOs and regulators, to encourage the creation and scaling of RECs, notably by delegating the operational management of their activities to a third-party service provider. Regarding energy sales, RECs do not need a supplier license unless their members mandate that they sell excess electricity beyond the defined local perimeter. The legislation also defines the role of "network managers," who are responsible for implementing the technical, administrative, and contractual conditions for REC operation, particularly in relation to electricity metering, under regulated tariffs. A DSO may be mandated to manage a closed distribution network on behalf of an REC. Tariff structures for grid use will

[3]For more information: Government of Wallonia Decree—17 March 2023–1. Qu'est ce qu'un bâtiment? - Site énergie du Service public de Wallonie, https://wallex.wallonie.be/eli/arrete/2023/03/17/2023044651/2023/10/08, Acessed: 03/06/2025

Table 5.1 List of all the active ECs in Wallonia according to CWaPE database

Name	Type	City	Date of approval	Energy-sharing authorization
uni BE 1 ASBL	CEC	Mons	17-03-2025	N/A
DURBUY 1 ASBL	REC	Durbuy	21-02-2025	N/A
Energie Libre ASBL	CEC	Gesves	21-10-2024	(CWAPE 2025a)
Soleil d'Aubange ASBL (CERSA)	REC	Aubange	28-05-2024	(CWAPE 2024)

be determined by the regulator, taking into account the benefits of the REC, such as avoided network losses, deferred infrastructure investments, and enhanced renewable energy deployment. However, REC participants are excluded from net metering, which remains available only to RSCs. Table 5.1 provides a detailed overview of the currently active ECs in Wallonia.

The list above (Table 5.1) shows the ECs that have the legal form of a non-profit association; however, there are 21 member cooperatives in Wallonia under "REScoop Wallonie"—comprising 20 full members and 1 associate member, contributing to the intensive efforts toward the ECs' creation and development.

"RESCoop Wallonie" is the federation that manages all renewable energy cooperatives in the region, which produce energy using numerous technologies, from PV and wind to hydroelectric and even biomethanization. The Walloon RESCoops consolidated their efforts into establishing COCITER ("Comptoir Citoyen des Energies") the only supplier for green energy in the region, already achieved a production of 117 million kWh, sufficient to meet the energy demand of 32,470 households. Initially, COCITER was founded by only 3 cooperatives, but later the other ones joined, it officially obtained its electricity supply license in 2015.[4]

Facilitators—comprising companies, intercommunal agencies, innovation networks, project developers, and strategic partners—play a critical role in accelerating the regional energy transition by providing technical, legal, financial, and organizational support. A key player is IDETA (Intercommunale de Développement Économique des Arrondissements de Tournai, d'Ath et de Communes Avoisinantes), a public development agency active in the Walloon region, supporting territorial sustainability and renewable energy initiatives, including wind energy projects, and facilitating the spatial and technical frameworks essential for EC development.

Complementing these efforts, IIS CETWA (Communautés d'Énergie et Technologie WAllonnes), established in 2024 under Wallonia's Smart Specialisation Strategy, acts as a collaborative innovation network. It unites companies, research institutions, universities, and associations focused on energy communities and technologies and fosters synergies, by promoting research infrastructure sharing and enhancing the valorization of R&D outcomes to strengthen Wallonia's presence in the sustainable energy sector. Together, these organizations contribute to a

[4] For more information: La Coopérative d'Electricité verte, locale et citoyenne, https://www.cociter.be/, Accessed: 03/06/2025

supportive ecosystem that advances the deployment and scaling of renewable energy communities throughout Wallonia (SPW. Stratégie de Spécialisation Intelligente 2022).

Under Article 27 of the Electricity Decree, CWaPE is eligible to authorize pilot projects to test alternative networks, new technologies, or tariff models for the Walloon energy market. Before a legal framework for energy communities, energy sharing, and self-consumption was in place, CWaPE approved several pilot projects to help identify best practices, benefits, and barriers, guiding future legislation and regulation; some of them are recorded in the list below:

- **SeaCoop Cooperative** (European Commission 2022): Founded in 2022 as a collective investment independent energy project of 34 renewable energy cooperatives (members of REScoop Flanders or Wallonia), collective energy actions and crowdfunding aimed the 20–20 approach for the supply of 20% of electricity generated by wind potential for covering the energy needs of Belgian households and SMEs through local suppliers, as Ecopower and COCITER with the citizen participation to be in the heart of the endeavor. In line with the business model developed, SeaCoop follows the social targets with a non-profit organization in an inclusive and democratic design (one member one vote) involving each member in the decision-making processes. Despite its innovative perspectives, SeaCoop faces a series of operational, technical, and societal challenges, especially in competing with established multinational consortia, which remain technically complex, while administrative issues related to contracting and a lack of gender balance are also recorded (European Commission 2022).
- **ZuidtrAnt (Mortsel)** (European Commission 2023c): Founded in 2016, ZuidtrAnt community is a cooperative of natural persons, local authorities, government agencies, NGOs, and associations, but also social housing companies, and private sector involvement counting 650 members, to develop activities around the electricity generation, monitoring, and physical renovations of the area's building stock. Various challenges are recorded for this case, especially on the lack of clarified regulatory frameworks, or financial support.
- **ACRus Project** (IDETA. Agence de Dévéloppement Territorial 2025): Conducted between April 2022 and March 2025 in Péruwelz, aimed to test two incentive-based network tariff models to encourage the sharing of solar energy within the same business building. Overall, the project managed a rate of 45% of self-consumption with a coverage rate of 37% of a persistent surplus and demonstrated the importance of energy sharing within a tertiary building, but also the networks' limits, critical for optimization problems through larger scales, as the community.
- **Ecopower** (2025): Prominent cooperative based in Flanders (founded in 1991), with focused activities on renewable production and distribution. Model for ECs' promotion with diverse objectives: electricity production from RES, promotion of energy efficiency, encouragement and support for democratic action plans, and involvement of fledgling partners in ECs' structures. Counting more than 70,000 members (2023), and various projects across Flanders, the cooperative

invests in PV installations on residential, commercial, and public buildings, often in collaboration with public authorities, exploring perspectives for other renewable resources to optimize collective benefits of self-consumption schemes committed to decentralized and sustainable energy models, aligned with European overall commitments, offering growth opportunities for ECs development in Belgian territories (Laurel 2024). Today, Ecopower serves 2% of the Belgian energy market and produces more than 100 million kWh using solar, wind power, and hydroelectricity (Ecopower 2021).
- **Courant d'Air** (2009): Citizen cooperative developed in the Walloon region, recognized as a social economy enterprise for the promotion of renewable energy. Founded in 2009 with a mission to develop local renewable generation and support autonomy through citizen participation projects and energy justice and democratization, the cooperative focuses on community engagement and learning activities to provide new participants and actors with tools for best practices in green electricity production. Comprising more than 4800 members with equal voting participation rights to all relevant activities (Boulanger et al. 2021), operates six (6) wind turbine collaborative projects, and biomass initiatives in its energy portfolio, aligned with EU engagements, and benefits from the Walloon Energy Decree framework for the ECs' definitions and operations, but also being supported by the EU rules and self-consumption promotion.

5.7 Unveiling Energy Communities Cases in Belgium

5.7.1 HospiGREEN: A Case of an Energy-Sharing Community in Wallonia

The project was implemented in Tournai (Belgium) from 2020 to 2023, involved ten (10) participants, mostly healthcare facilities (e.g., hospitals and retirement homes). This case is *the first of its kind* REC in Wallonia region, according to the EU relevant Directive (European Commission 2019) for decentralized energy management and renewable energy integration and motivated by various types of projects, e.g., wind turbines, photovoltaic installations, and so on for the self-consumption maximization, and increase in local generation and use (Viadere 2025).

5.7.2 ZuidtrAnt as a Flemish EC Case Study

The cooperative strives to offer affordable, sustainable energy by utilizing investments from citizen-purchased shares. These funds support renewable energy projects, with a team of multidisciplinary experts applying specialized knowledge to bring these initiatives to life. Operating primarily in and around Antwerp, ZuidtrAnt

Table 5.2 Projects of ZuidtrAnt as CEC[a]

Projects	Year	Progress
Projects sun	2023	9 new solar roofs, with a capacity of 644 kWp
		28 solar roofs with a full capacity 1656 MWh to supply 382 household
SharedSun project	2023	Annual saving of 272 tons of CO_2
		Combined solar roofs with electric shared mobility
TANDEMS	–	Seven solar roofs, nine charging stations and 9 shared cars
RE-FLEX	2023	Synergies of ECs with local actors
		Scalable and modular charging plaza concept
		Renewable energy production with energy
		Storage in both stationary and EV batteries
KlimaatWerf	2023	154 solar installations
		90 energy renovations

[a]For more information: https://www.zuidtrant.be/projecten, Accessed on: 03/06/2025

focuses on enhancing energy efficiency across its projects to speed up the transition to renewable energy (Table 5.2).

5.7.3 *Hassrode (Leuven): A Case of a Business Park*

The Hassrode business park primarily includes residential functions, service-oriented activities, and significant investments in photovoltaics (Lode et al. 2022). Almost no renewable energy projects have yet started up, and investments in EC incentives are in the early stages (Heuninckx et al. 2022).

5.7.4 *Buurzame Stroom (Ghent): A Case of a Residential City Neighborhood*

A grassroots initiative for the integration of sustainable energies at the district level of Ghent-Dampoort case in the northwest part of Belgium. During EC development, a citizen energy cooperative, "Energent," was established, including members of the participating residential dwellings, and brought together different stakeholders (20 prosumers from a 36-building household). The main goal of the initiative has been to increase the PV investments and relevant projects in an inclusive participation design (Lode et al. 2022).

5.7.5 Green Energy Park (Zellik, near Brussels): A Case of a Commercial City Neighborhood

Situated in the city of Zellik, near Brussels, the district comprises an existing industrial zone with 21 commercial buildings and a planned innovation center. In terms of equipment, the district interconnects prosumers in incubator start-ups and a large parking lot equipped with electric vehicles, aiming to be further expanded and integrate other relevant infrastructure of solar panels, wind turbine, cogeneration, and energy storage facilities (Lode et al. 2022).

5.8 Benchmarking Analysis of Belgian Case Studies

See Table 5.3.

5.9 Key Observations of Energy Communities Projects of Belgium

Belgium has seen significant growth in energy-sharing initiatives; however, this development varies widely across regions due to region-specific factors, also differ by region. Wallonia has a high number of cooperatives, whereas Flanders has a larger share of RSCs, and Brussels has a tendency to favor CSCs model.

In Belgium, each of the three federations has adopted its unique approach to transposing RSC, REC, and CEC definitions at designated timelines. Flanders has drafted and adopted a legislative framework in December 2020 for all three mentioned types of ECs. As for Wallonia, collective self-consumption had decreased in 2018 and 2019 with legal framework defined for REC definition and absence for CECs. On its side, Brussels has adopted the 2018 definition for Collective Self-Consumption and transposed the EU definitions for REC and CEC by 2021.

Each Belgian region has established its own procedures for setting up ECs.

A lack of initiative from local stakeholders, supporting mechanisms, and associated institutional and legal constraints are associated with structural obstacles to energy community applications. Nonetheless, probably the most important challenge in Belgium remains the *regional government structure with a heterogeneous legal framework foreseeing diverse directives for the three federations* (ROLECS 2020).

Looking specifically at energy community barriers, the analysis for the Belgian frame identifies a *lack of experience and expertise in the community models* at social, cultural, and organizational levels. The challenges are multiple, starting with *transparency and trust by the prosumers, the complicated legal and funding support schemes, and the lack of previous knowledge of relevant successful stories.*

Table 5.3 Benchmarking of EC development in the federations of Belgium

Criteria	Wallonia	Brussels-Capital Region	Flanders
Energy regulator	CWaPE (Commission Wallonne Pour Energie) (CWAPE 2019)	Sibelga—DSO Brugel—Responsible for authorizing and validating EC permission.	VREG
Supply license and supplier obligations	Based on capacity, geographical area, and type of consumers	Based on type of electricity supplied, capacity, and type of consumers	Only one type of supply license
The role of DSO	Granted right for RECs to become DSOs	Granted right for RECs to become DSOs	Granted right for RECs to become DSOs
Membership conditions	N/A	N/A	Only if an EC is not a supplier, any fee or notice period for leaving EC
Management of ECs	Third party as manager and be sole point of contact with CWaPe and DSO	N/A	N/A
Legal form	The legislator allows ECs to have the freedom to regulate themselves in the way they want, but impose minimum conditions by law	N/A	N/A
Legislative frameworks	Decree of 5 May 2022 AGW of 17 March 2023	Energy decree and VREG guidelines	Ordinance of 2021 and specific classification via Brugel
Licensing procedures	Procedure clear and rigorous, CWaPE is responsible	Clear and well-structured licensing procedures in a long time for verification and approval, it can take up to a year (Basem 2022)	No well-defined criteria for who can register, and VREG does not oversee to ensure compliance
EC facilitators	IDETA RESCoop Wallonie	Sibelga Bruxelles Environnement Energie commune	Energiehuis (EnergyHouse) REScoop Vlaanderen Technical Assistance Hubs (TEAH)
EC definitions adopted	Renewable Self-Consumption (RSC) CSC REC and CEC	P2P energy trading CSC LEC, REC, CEC	Renewable Self-Consumption (RSC) P2P energy trading CSC REC and CEC
Common legal forms	Cooperative Society (SC) Non-profit Association (ASBL)	Cooperative Society (SC) Non-profit Association (ASBL)	Cooperative Society (SC) Non-profit Association (ASBL)

(continued)

Table 5.3 (continued)

Criteria	Wallonia	Brussels-Capital Region	Flanders
Consumer protection	N/A	N/A	If an EC is a supplier; it is obliged to provide simple and fair internal complaints system
Scope of proximity	The Wallon legislator opted for a hybrid version of geographical and technical criteria	N/A	N/A
Effective control	Controlled by the participants located near the production. Proximity is defined in terms of eligibility (RESCOOP.EU 2022)	Members close to the projects developed by the community (Sibelga 2025)	CECs and RECs are required to elaborate in their articles of association the rules governing control
	Natural persons, local authorities as defined by the government, including municipalities and small businesses (RESCOOP.EU 2022)	Exclusive to natural people, local authorities and small enterprises that do not use it as their main business or professional activity	ECs and RECs are instead required to elaborate in their articles of association the rules governing control
Activities	Electricity production, supply Self-consumption Sharing, via the public distribution or local transmission network, Aggregation Participation in the provision of flexibility services Energy storage Charging services for electric vehicles Sale of self-produced electricity that is neither self-consumed nor shared (SPW. Stratégie de Spécialisation Intelligente 2022)	Store energy Engage in energy efficiency actions Supply electricity to third parties (*subject to a supply license*) Contribute to demand-side flexibility Offer electric vehicle charging solutions (environnement.Brussels 2025)	Produce energy from installation, whether directly Self-consumption of the energy referred to in point 1 Store energy Flexibility or participant in flexibility or aggregation Charging services for electric vehicles Energy sharing, between the partners or members, of the energy referred to in point 1, in accordance with Article 7.2.1, second paragraph (Flemish Codex 2025)

5.9.1 Flanders

Flanders stands out as the most advanced region in Belgium in terms of both the quantity and variety of energy-sharing models. The region has effectively transposed EU Directives, particularly those concerning RECs and CECs, into its regulatory framework along with a high concentration of active projects and the presence of 24 well-established Renewable Energy Source Cooperatives (REScoops), demonstrating strong support for citizen participation and decentralized renewable energy production (REScoop Valaanderen 2026).

Overall, Flanders benefits favorable regulatory, institutional, and economic conditions that promote the growth of energy communities. A clear example of this supportive environment is Ecopower (see Sect. 5.6). By 2022, the cooperative operated 20 wind turbines, one hydropower plant, one cogeneration facility, a district heating network, and around 250 decentralized solar PV installations on the rooftops of schools and homes across Flanders. Ultimately, within the same year, it produced 74 million kWh of green electricity, resulting in 61,000 tons of CO_2 emissions avoided (Ecopower 2025).

5.10 Brussels-Capital Region

Brupower is a representative energy cooperative in the Brussels-Capital Region, exemplifying citizen-led initiatives within an urban context. Established in May 2022, the cooperative brings together local residents to collectively invest in and co-own renewable energy projects, with a primary focus on solar photovoltaic installations deploying approximately 4500 solar panels across Brussels in the coming years. Brupower's approach reflects the region's emphasis on combining energy transition with social inclusion. Its first project, implemented in partnership with DoucheFLUX—an NGO assisting homeless individuals—provides renewable electricity to power essential services, such as showers, laundry facilities, medical support, and legal assistance. This demonstrates how energy cooperatives in Brussels often merge environmental goals with social solidarity. Currently, Brupower is supported by more than 40 volunteers, who contribute their time and expertise in fields such as energy regulation, communication, citizen engagement, photovoltaic project management, and digital tools. As a representative cooperative in Brussels, Brupower illustrates the potential of community-driven energy models to foster both sustainability and social impact in urban settings (brupower 2025).

5.11 Wallonia

The Decree of the Walloon Region transposed the CEP definitions of REC and CEC into its legislation in 2018, followed by a transposition of the REC definitions in the decree law in 2019, and entered into force in 2022 (Wallex 2023).

To support the efficient and careful development of RECs in Wallonia, five (5) pilot projects have been conducted. HospiGREEN project, the first Renewable Energy Community registered in Wallonia, is concentrated on the promotion of local energy generation and decentralized management objectives (CWAPE 2025b) motivated by local actors (e.g., Luminus) for a stable rent with less imbalances from surplus generation (Viadere 2025) involved ten (10) participants (Viadere 2025).

The largest Walloon energy cooperative is COCITER ("Le Comptoir Citoyen des Énergies"), a cooperative that supplies green electricity produced by its RESCoop members in the region. Founded in 2012 by three Walloon citizen energy cooperatives, COCITER was established to provide locally generated renewable electricity directly to citizens. As of 2023, it brings together 15 associated cooperatives and serves around 10,000 clients, covering the electricity needs of approximately 16,000 households through a diversified energy mix that includes wind, solar, hydro, and biomethane sources. COCITER functions as a non-profit entity, reinvesting its revenues into operations and community projects, with the core aim of strengthening citizen ownership and decentralizing energy production in Wallonia (COCITER 2025).

5.12 Gaps and Barriers in Belgian Energy Communities: What to Improve?

In Belgium, ECs' growth is uneven due to fragmented regional regulations and a lack of central oversight, as each federation sets its own rules, leading to confusion, uncertainty, and a lack of trust in future EC development. Flanders leads with strong legal support and active citizen cooperatives, while Wallonia struggles with unclear and changing legislation and limited funding. Brussels, though slower in projects, excels in transparency through Brugel's monitoring platform.

Nevertheless, despite regional differences, all regions have seen an increase in community-led energy initiatives driven by supportive policies, e.g., feed-in tariffs or tax incentives, encouraged by the cooperatives' organization and operation, such as Ecopower and Courant d'Air, with interesting progress, however, varied across the regions.

Belgium highlights gaps in its national registry, leading to unclear oversight, unequal regional development, legal uncertainty, limited access to funding, and, more importantly, missed opportunities for collaboration, which could hinder the country's energy transition. Cooperatives often scale beyond local RECs, promoting fairer markets and clean energy.

Given the prevailing desynchronization among the Belgian federations, a unified national strategy is simply not an optimal solution, as regional contexts demand their own interventions. Flanders can expand under its current framework, while Wallonia and Brussels need simplified procedures, technical support, and more targeted funding to support the integration of ECs in addressing energy poverty, as the inclusion of low-income households ought to be one of the key objectives for setting up EC, yet it remains unclear how it addresses these groups in their EC implementation. *Harmonized definitions*, based on EU categories (RECs and CECs), are essential for clearer regulation and communication across regions. Furthermore, *interregional collaboration and information* sharing are a promising pathway to more coordinated and effective EC development.

References

S. Basem. *Already 200 Proposals for a Brussels Energy Community.* (2022). https://brusselsmorning.com/already-200-proposals-for-a-brussels-energy-community/26144/

S. Boulanger, M. Massari, D. Longo, B. Turillazzi, C.A. Nucci, Designing collaborative energy communities: a European overview. Energies **14** (2021)

Brugel. Le regulateur bruxellois pour l'énergie (2025) *Observatoire du partage d'énergie en région de Bruxelles-Capitale 2022–2025.* https://app.powerbi.com/view?r=eyJrIjoiMTRiMzFlOTQtODgwNi00OTk5LTk5YTYtMzU1YTE0NjQ1YTRmIiwidCI6ImMwYjg2YzA3LWRhZGUtNDkyMC1hYzEzLWIwZWNhZDNiMmM5NSIsImMiOjh9

brupower (2025) *brupower. La coopérative d'énergie citoyenne bruxelloise.* https://coop.brupower.be/fr_BE

Bruxelles environnement. (2023). *Partge d'électricité au sein d'un même bâtiment. Facilitateur "Partage et Commu".* https://environnement.brussels/media/12774/download?inline

COCITER. *COCITER. Electricité coopérative.* (2025). https://www.cociter.be/

Courant d'Air, COCITER Electricité Coopérative. Courant d'Air. (2009) https://www.cociter.be/project/courant-dair/

CWAPE. *Communautés d'énergie.* https://www.cwape.be/node/158 (2019)

CWAPE. *Décret du 5 mai 2022 modifiant diverses dispositions en matière d'énergie dans le cadre de la transposition partielle des directives européennes du 'Clean Energy Package'.* https://www.cwape.be/node/5111. (2022)

CWAPE. *Décision CD-24g11-CWaPE-0958 autorisant une activité de partage d'énergie-Communauté d'energie renouvelable 'Soleil d'Aubange'.* https://www.cwape.be/sites/default/files/cwape-documents/2024.07.11-0958-Decision_autorisation_partage_PA0001_CER_Soleil_Aubange.pdf. (2024)

CWAPE. *Décision CD-25a30-CWaPE-1039 autorisant une activité de partage d'énergie-Communauté d'Energie Citoyenne 'Energie Libre'.* https://www.cwape.be/sites/default/files/cwape-documents/2025.01.30-1039-Decision_autorisation_partage_PA0003_Energie%20Libre.pdf. (2025a)

CWAPE. *Evaluation du cadre relatif aux Communautés d'Energie au partage d'énergie et à l'autoconsommation.* https://www.cwape.be/sites/default/files/cwape-documents/2025.02.20-0109-Rapport%20%C3%A9valuation%20CE%20et%20partage.pdf. (2025b)

EAS. Energy Annual Statistics. (2010). *Annual Statistics.* https://pure.mpg.de/pubman/faces/ViewItemOverviewPage.jsp?itemId=item_1164645

Ecopower. *Ecopower. Annual report.* (2021)

Ecopower. *Sustainable energy from, for and by citizens.* (2025). https://www.ecopower.be/

environnement.brussels. *Partage et Communautés d'énergie.* (2025). https://environnement.brussels/citoyen/lenvironnement-bruxelles/agir-eco-responsable/partage-et-communautes-denergie

European Commission. *Regulation (EU) 2019/943 of the European Parliament and of the Council of 5 June 2019 on the internal market for electricity (recast).* (2019). https://eur-lex.europa.eu/eli/reg/2019/943/oj/eng

European Commission, *SeaCoop* (Energy Communities Repository, Berchem, Belgium Energy Community, 2022) https://europa.eu/assets/wcloud/widgets/202406/c8c73fd0-2f18-11ef-94d0-599e9b3ca11b/27946.pdf

European Commission, *Energy Communities Repository* (Overview of the policy framework, Belgium (Brussels Capital Region), 2023a) https://circabc.europa.eu/ui/group/8f5f9424-a7ef-4dbf-b914-1af1d12ff5d2/library/1b0260b9-76e0-4eea-9330-f10907d19cc6/details?download=true

European Commission. (2023b). Renewable energy directive (RED II) directive (EU) 2018/2001 (recast) on the promotion of the use of energy from renewable sources.. https://www.europex.org/eulegislation/renewable-energy-energy-directive/

European Commission, *ZuidtrAnt* (Mortsel, Belgium, Energy Communities Repository, 2023c) https://europa.eu/assets/wcloud/widgets/202406/c8c73fd0-2f18-11ef-94d0-599e9b3ca11b/27749.pdf

European Commission. *Regulation (EU) 2024/1747 of the European Parliament and of the Council of 13 June 2024 amending Regulations (EU) 2019/942 and (EU) 2019/943 as regards improving the Union's electricity market design.* (2024). https://eur-lex.europa.eu/eli/reg/2024/1747/oj/eng

A. Felice, L. Rakocevic, L. Peeters, M. Messagie, T. Coosemans, L.R. Camargo, An assessment of operational economic benefits of renewable energy communities in Belgium. J. Phys. Conf. Ser. **2042**(1), 012033 (2021). https://doi.org/10.1088/1742-6596/2042/1/012033

Flanders. *Call green electricity.* (2021). https://www.vlaanderen.be/call-groene-stroom

Flemish Codex. *Decree laying down general provisions on energy policy.* (2025).. https://codex.vlaanderen.be/Zoeken/Document.aspx?DID=1018092&param=inhoud&AID=1291906

Flemish Government. *Decree of the Flemish Government Amending Regulations on Energy, ETS and the Building Pass. Legal Bases.* (2010)

D. Frieden, A. Tuerk, J. Roberts, S. D'Herbemont, A. Gubina. *Collective Self-Consumption and Energy Communities: Overview of Emerging Regulatory Approaches in Europe. Compile: Integrating Community Power in Energy Islands.* (2019). https://www.rescoop.eu/uploads/rescoop/downloads/COMPILE_Collective_self-consumption_EU.pdf

D. Frieden, A. Tuerk, C. Neumann, S. D'Herbemont, J. Roberts. *Compile. Collective Self-consumption and Energy Communities: Trends and Challenges in the Transposition of the EU Framework.* (2020). https://www.researchgate.net/publication/346975546_Collective_self-consumption_and_energy_communities_Trends_and_challenges_in_the_transposition_of_the_EU_framework?channel=doi&linkId=5fd72ab845851553a0b571ac&showFulltext=true

S. Heuninckx, G.T. Boveldt, C. Macharis, T. Coosemans, Stakeholder objectives for joining an energy community: Flemish case studies. Energy Policy **162**. Scopus (2022). https://doi.org/10.1016/j.enpol.2022.112808

IDETA. Agence de Dévéloppement Territorial. *Mise en oeuvre du projet pilote-Structure tarifaire réseau ACRus 'Auto Consumption in Real Estate for us'* [Report]. (2025). https://www.cwape.be/sites/default/files/cwape-documents/2025.06%20ACRus%20rapport%20projet%20VF.pdf

C. Inês, P.L. Guilherme, M.G. Esther, G. Swantje, H. Stephen, H. Lars, Regulatory challenges and opportunities for collective renewable energy prosumers in the EU. Energy Policy **138**, 111212 (2020)

M. Kumar Singh, S. Mahapatra, J. Teller, An analysis on energy efficiency initiatives in the building stock of Liège, Belgium. Energy Policy **62**, 729–741 (2013)

J. R. Laurel. *Regulatory frameworks and community engagement strategies: A comparison of Belgian policy against Denmark and The Netherlands* [Master thesis, UHasselt. Faculty of Business Economics]. (2024). https://documentserver.uhasselt.be/bitstream/1942/43971/1/6b65d81b-e0b1-4a26-a449-62fe7a69690d.pdf

M.L. Lode, S. Heuninckx, G. te Boveldt, C. Macharis, T. Coosemans, Designing successful energy communities: a comparison of seven pilots in Europe applying the multi-actor multi-criteria analysis. Energy Res. Social Sci. **90** (2022)

Région de Bruxelles Capitale, Plan énergie climat 2030. The right energy for your Region. (2019) https://document.environnement.brussels/opac_css/elecfile/plan_energie_climat_klimaat-plan_2030_fr

F. Renard, S. Nourricier, S. Pietrantonio, V. Feldheim, *Technical-Economic Analysis of the Cost-Effectiveness of Energy Saving Investments for Residential Buildings* (Service Public de Wallonie, DG04, 2008)

https://rescoopv.be/, Accessed on: 17/05/2026, RESCOOP Vlaanderen. One mission: more sustainable citizen energy

RESCOOP.EU. (2022). *Enabling Frameworks and Support Schemes. REC & CEC Definitions. Overall Assessment.* https://www.rescoop.eu/policy/transposition-tracker/rec-cec-definitions/brussels-rec-cec-definitions

RESCOOP.EU, *Enabling Frameworks and Support Schemes* (RESCOOP.EU, Belgium (Flanders), 2025a) https://www.rescoop.eu/policy/transposition-tracker/enabling-frameworks-support-schemes/belgium-flanders

RESCOOP.EU, *European Energy Communities Facility* (Belgium-Capital Region of Brussels. Energy Communities Policy Framework, 2025b) https://energycommunitiesfacility.eu/sites/default/files/policy-country-fitches/EECF%20country%20fiche%20Belgium%20-%20Brussels%20-%20update%20May%202025%20v2.pdf

ROLECS. (2020). *Legislative options and obstacles for energy communities in Belgium—Summary of key issues identified and recommendations. Task 2.3.1.* https://www.brussels-school.be/output/news/rolecs-report-identifies-key-legislative-challenges-energy-communities-belgium#:~:text=14.10.2020-,ROLECS%20report%20identifies%20key%20legislative%20challenges%20for%20energy%20communities%20in,energy%20and%20the%20energy%20transition

Service Public de Wallonie. Service Public Fédéral. Justice. *5 MAI 2022. - Décret modifiant diverses dispositions en matière d'énergie dans le cadre de la transposition partielle des directives 2019/944/UE du 5 juin 2019 concernant des règles communes pour le marché intérieur de l'électricité et 2018/2001/UE du 11 décembre 2018 relative à la promotion de l'utilisation de l'énergie produite à partir de sources renouvelables et en vue d'adapter les principes relatifs à la méthodologie tarifaire.* (2022). https://www.ejustice.just.fgov.be/cgi/article_body.pl?language=fr&caller=summary&pub_date=2022-10-05&numac=2022033591

Service Public de Wallonie. Service Public Fédéral. Justice. *17 MARS 2023. - Arrêté du Gouvernement wallon relatif aux communautés d'énergie et au partage d'énergie.* (2023). https://www.ejustice.just.fgov.be/cgi/article_body.pl?language=fr&caller=summary&pub_date=2023-09-28&numac=2023044651

Sibelga. *Sharing energy in a community.* (2025). https://www.sibelga.be/en/connections-meters/renewable-energy/energy-sharing/sharing-energy-in-an-energy-communitiy

SPW. Stratégie de Spécialisation Intelligente. *CETWA (Communautés d'Energie et Technologiques WAllonnes).* (2022)

E. Viadere, Promoting energy-sharing communities: Why and how? Lessons from a Belgian pilot project. Energy Policy **198**, 114483 (2025)

Wallex. (2023). *17 mars 2023—Arrêté du Gouvernement wallon relatif aux communautés d'énergie et au partage d'énergie.* https://wallex.wallonie.be/eli/arrete/2023/03/17/2023044651/2023/10/08

Walloon Parliament. (2014). *The New Walloon Climate Decree: stringent objectives towards effective greenhouse gas emissions reductions..* https://www.lexgo.be/fr/actualites-et-articles/2109-the-new-walloon-climate-decree-stringent-objectives-towards-effective-greenhouse-gas-emissions-reductions

M. Williams. *Belgian houses: beauty is in the eye of the brick holder.* (2023). https://ww3.rics.org/uk/en/modus/built-environment/homes-and-communities/belgian-house-design.html

Chapter 6
Discussion

6.1 Rethinking Business Models of Energy Communities Toward Social Innovation Actions

Most challenges in the operational and deployment activities of EC implementation are non-technological and related to business models and a lack of stakeholder engagement. In this regard, innovative prototypes and research pipelines, driven by strong leadership and powerful participation, are required to create robust governance ecosystems and achieve integrated urban transformation, acceptance, and replicability. In fact, a community is a complex, scalable operation capable of supporting transition models and improving users' quality of life.

In practice, the implementation of the concept applies to a community-level approach, driven by shared thinking, to develop efficient synergies and leverage diverse skills to realize the social potential of decentralized energy solutions and empower local communities through collective action for shared benefits. To do so, a redesign of existing business models converted into systemic, alternative, and dynamic methods seems necessary.

Having its own set of complexities and circumstances, existing business models cover engineering design, including technological dimensions of their energy efficiency deployments from their installation, operation, and monitoring, e.g., demand-response flexibility. Today, as energy is at the core of European and national agendas, rethinking alternative business models to establish cooperative models and invest in sustainable transitions is essential to build trust. Adopting approaches to promote synergistic development and new forms of partnerships is a key operational focus for future design, going beyond technical conceptions. Urban transformation requires a well-developed roadmap, in which digitalization and virtual environments will be the communication channels during all the operational stages to facilitate interactions and collaboration between all stakeholders and enable decision-making capabilities.

S. Koutra, *Energy in the Community*, SpringerBriefs in Applied Sciences and Technology, https://doi.org/10.1007/978-3-032-27148-8_6

Nevertheless, emerging new business models include new types of activities, values, and governance forms, e.g.., Nolden et al. (2020), shaping differences in physical, technical, operational conditions, and mechanisms. Moroni and Tricarico (2018) argued on the understanding of the community in an organizational system "non-ideological," where models as peer-to-peer or microgrids are all considered local productions and forms of communities, e.g., Koirala et al. (2016).

Active citizen participation and engagement are assumed to produce multiple outcomes for community operations success stories and enable grassroots and citizen-led social innovations (Bielig et al. 2022). Caramizaru et al. (Caramizaru and Uihlein 2020) define *ECs as a type of social innovation* that promotes energy justice and prosumers, with democratic decision-making processes placed in users' hands through participatory approaches. As top-down and bottom-up approaches, they are supposed to strengthen energy justice (van Bommel and Höffken 2021) and democracy even though, despite the aforementioned benefits, there is a lack of evidence of the interrelationships of the functioning of how ECs maximize collective benefits, e.g., Berka and Creamer (2018) and Moroni et al. (2019), usually characterized as "romanticized narratives" (Hanke et al. 2021).

Although the social impact goes beyond purely economic dimensions, encompassing significant transformations in social, cultural, and community realities, this suggests that the scope of voluntary or third-sector actions extends beyond the financial to address fundamental aspects of well-being and human development. Vanclay et al. (2003) focus their work on a two-level approach (individual and community) across physical and perceptual dimensions, since social impact can be physical or perceptual and emotional.

6.2 Energy Community. An Approach Toward Social Innovation?

Although considerable documentation on the energy community seems to exist, there is still a wide interpretation of the term, evolving over time, seeking open and participatory processes, outcomes, and collective benefits, locally based (Devine-Wright 2019), while others have a broader organizational structure, e.g., Moroni and Tricarico (2018), by placing more emphasis on the outcomes rather than the processes but also on other spectrums opposed to only local incentives, e.g., Hicks and Ison (2018).

Traditional studies have devoted considerable attention to understanding how cooperatives effectively promote clean energy in heterogeneous settings. In fact, cooperatives have a critical role in the transition dynamics, e.g., Oteman et al. (2017) and in energy democratization, expanding the adoption and acceptance of energy practices and establishing institutional cooperation. Adopting multilevel perspectives to analyze the energy transition socio-technically, as a systemic and interactive process, provides a valuable conceptual framework by dividing the spatial and building contexts by focusing on local embeddedness (Coenen and Hoppe 2022). As downstream experiences, ECs are assumed to foster higher acceptance of energy technologies, thereby promoting cohesion, capital, and community empowerment, e.g., Coy et al. (2021), but, with marginalized limitations of operationalized participation, may achieve social and energy transformative actions.

So far, policymakers and practitioners have placed significant importance on the EC as a tool to accelerate the energy transition, often overlooking the social complexities of this transformation at the national and local levels, especially regarding the engagement and inclusion of multiple actors.

Overall, transitions are described as multi-actor developments with complex interactions (Grin et al. 2010) and ambiguities on roles and responsibilities streaming into decision-making at different levels of structured transitions in relation to the bottom-up energy landscape and polycentric and multilevel governance, and varied technical learning, e.g., cognitive, normative, or relational, demanding for transformative and adaptive design for transposing the nomenclature of the status quo patterns.

Energy communities are a case of socio-technical innovation and governance arrangements, as discussed in the existing literature, including socio-technical thinking. As a social innovation, ECs reconfigure social and interactive practices by questioning existing systems (Hiteva and Sovacool 2017) and employing bottom-up approaches, where citizen participation is at the core of the business models, changing the status quo of technocratic solutions and promoting justice and democracy occurring in energy and social transformations, operating as a "social contract" in an open and voluntary processes, including a broad range of actors (e.g., natural persons, local authorities, private enterprises, etc.).

In recent years, various viewpoints have been presented in line with the involvement of ECs in enhancing sustainability, energy independence, and community resilience, by creating bidirectional activities between stakeholders in the direction of optimal distribution, flexibility, and stability of energy systems, but also in line with decentralization challenges, alleviation of environmental risks, improvement of social welfare, and energy hybridization for altering carbon systems trajectories in the years to come.

In its "citizen" governance form, the term "energy community" operates as an organizational structure through collective actions and bottom-up rules. Broadly, membership is open and voluntary, including a wide range of actors, such as natural persons, local authorities, and small enterprises, in legal entity status.

References

A.L. Berka, E. Creamer, Taking stock of the local impacts of community owned renewable energy: A review and research agenda. Renew. Sust. Energ. Rev. **82**, 3400–3419 (2018)

M. Bielig, C. Kacperski, F. Kutzner, S. Klingert, Evidence behind the narrative: critically reviewing the social impact of energy communities in Europe. Energy Res. Soc. Sci. **94** (2022)

A. Caramizaru, A. Uihlein, *Energy Communities: An Overview of Energy and Social Innovation* (JRC Science for Policy Report, European Commission, 2020) https://publications.jrc.ec.europa.eu/repository/handle/JRC119433

F.H. Coenen, T. Hoppe, *Renewable Energy Communities and the Low Carbon Energy Transition in Europe* (Springer, 2022)

D. Coy, S. Malekpour, A.K. Saeri, R. Dargaville, Rethinking community empowerment in the energy transformation: a critical review of the definitions, drivers and outcomes. Energy Res. Social Sci. **72** (2021)

P. Devine-Wright, Community versus local energy in a context of climate emergency. Nat. Energy **4**, 894–896 (2019)

J. Grin, J. Rotmans, J. Schot, Transitions to sustainable development. New Directions (2010) (Taylor & Francis)

F. Hanke, R. Guyet, M. Feenstra, Do renewable energy communities deliver energy justice? Exploring insights from 71 European cases. Energy Res. Social Sci. **80** (2021)

J. Hicks, N. Ison, An exploration of the boundaries of 'community' in community renewable energy projects: navigating between motivations and context. Energy Policy **113**, 523–534 (2018)

R. Hiteva, B. Sovacool, Harnessing social innovation for energy justice: a business model perspective. Energy Policy **107**, 631–639 (2017)

B.P. Koirala, E. Koliou, J. Friege, R.A. Hakvoort, P.M. Herder, Energetic communities for community energy: a review of key issues and trends shaping integrated community energy systems. Renew. Sust. Energ. Rev. **56**, 722 (2016)

S. Moroni, L. Tricarico, Distributed energy production in a polycentric scenario: policy reforms and community management. J. Environ. Manag **61**, 1973 (2018)

S. Moroni, V. Alberti, V. Antoniucci, A. Bisello, Energy communities in the transition to a low-carbon future: a taxonomical approach and some policy dilemmas. J. Environ. Manag. **236**, 45 (2019)

C. Nolden, J. Barnes, J. Nicholls, Community energy business model evolution: A review of solar photovoltaic developments in England. Renew. Sust. Energ. Rev. **122**, 109722 (2020)

M. Oteman, H.J. Kooij, M.A. Wiering, Pioneering renewable energy in an economic energy policy system: the history and development of Dutch grassroots initiatives. Sustainability **9** (2017)

N. van Bommel, J.I. Höffken, Energy justice within, between and beyond European Community Energy initiatives: a review. Energy Res. Social Sci. **79** (2021)

F. Vanclay, International principles for social impact assessment. Impact Assess. Proj. Apprais. **21**, 5–12 (2003)

Chapter 7
Energy Community: Which Future and How to Overcome Barriers and Ensure Replicability?

In previous sections of this book, we shed light on the multiple barriers and complexities of EC operationalization. Aspiring ECs mean increasing funding capacity and making upfront investments while overcoming barriers in the legal framework, social resistance, and technical difficulties. In some countries, efforts to mitigate technological and operational obstacles are hindered by a lack of digital infrastructure and insufficient dissemination of knowledge about alternative solutions in local contexts. Democratization of the electricity market and technical autonomy are key drivers of stakeholder and citizen engagement, delivering collective benefits in line with the principle of "no one left behind" and support for marginalized or vulnerable population groups. Simultaneously, the resilience of decentralized energy systems to mitigate technical disruptions can make an important contribution to demand-related models and flexibility, thereby increasing capacity and supporting congestion alleviation.

At the same time, clear, simplified, and unified definitions in both legal documentation and the overall transposition of the relevant European Directives are requested to reduce uncertainties and hesitations regarding the EC's establishment. Ensuring accessibility and knowledge dissemination are key factors to eliminate hindrances to the EC deployment; at the time being, few citizens have prior knowledge for the participation processes and the EC establishment; to mitigate these risks, it is important to counterbalance the ambiguities and provide relevant actors with explicit and easily comprehensible frameworks with best practices' share through handbooks, pamphlets, and relevant guidance.

Community engagement seems to be the key and driver for EC development and growth. Active involvement of all relevant stakeholders, in all forms, at every stage of the project, from planning through monitoring and management, is paramount. Local ownership and management control are emphasized in the literature and real-time settings through cooperatives for trust and transparency development, while capacity building, continuous learning, and dissemination are integral dimensions for the knowledge promotion and transferability to all involved parties including

S. Koutra, *Energy in the Community*, SpringerBriefs in Applied Sciences and Technology, https://doi.org/10.1007/978-3-032-27148-8_7

risk and impact assessment methods, essential to foster accurate progress and enable feedback.

Along with this, an important prerequisite for the EC deployment is the community acceptance and the collective dynamics to enable multi-stakeholder participation and co-design contributions toward the energy transition roadmap, involving actors in setup processes, as key drivers of addressing social dimensions in these incentives are outlined as effective practices for knowledge acquisition, sharing, development, and innovation, ensuring apart from the technical, the social performance as well from a viewpoint of satisfying the local needs by the daily roll-out of ECs within commonly accepted standards, and processes, in which all relevant actors are involved from the beginning of the EC's conceptualization and planning processes.

Along similar lines, capacity building and education are important levers for empowering community members and enabling their participation in developing sustainable energy systems. In this context, various forms can be observed, for instance, the technical or policy training or skills development for decision-making processes and new partnership (synergies) schemes. Other essential tools for further EC development are communication methods and information, including relevant digitalized and traditional tools, e.g. platforms, forums, etc., for raising awareness to a broad audience and sharing best practices' information.

MIX
Papier aus verantwortungsvollen Quellen
Paper from responsible sources
FSC® C105338

If you have any concerns about our products,
you can contact us on
ProductSafety@springernature.com

In case Publisher is established outside the EU,
the EU authorized representative is:
Springer Nature Customer Service Center GmbH
Europaplatz 3, 69115 Heidelberg, Germany

Printed by Libri Plureos GmbH
in Hamburg, Germany